Microwave Materials Handbook

Microwave Materials Handbook

Edited by **Alessandro Torello**

New York

Published by NY Research Press,
23 West, 55th Street, Suite 816,
New York, NY 10019, USA
www.nyresearchpress.com

Microwave Materials Handbook
Edited by Alessandro Torello

International Standard Book Number: 978-1-63238-329-7 (Hardback)

Contents

Preface

This book was inspired by the evolution of our times; to answer the curiosity of inquisitive minds. Many developments have occurred across the globe in the recent past which has transformed the progress in the field.

The book aims to serve as a resource guide for microwave materials. It discusses current studies on fundamental and modern measurement methods for the evaluation of substances at microwave frequencies. The book is separated into several different sections. It deals with current contributions on modern methods for the evaluation of dielectric substances. Microwave evaluation of organic tissues is, also, thoroughly explained. This book intends to provide useful information regarding the above stated topic to its readers.

This book was developed from a mere concept to drafts to chapters and finally compiled together as a complete text to benefit the readers across all nations. To ensure the quality of the content we instilled two significant steps in our procedure. The first was to appoint an editorial team that would verify the data and statistics provided in the book and also select the most appropriate and valuable contributions from the plentiful contributions we received from authors worldwide. The next step was to appoint an expert of the topic as the Editor-in-Chief, who would head the project and finally make the necessary amendments and modifications to make the text reader-friendly. I was then commissioned to examine all the material to present the topics in the most comprehensible and productive format.

I would like to take this opportunity to thank all the contributing authors who were supportive enough to contribute their time and knowledge to this project. I also wish to convey my regards to my family who have been extremely supportive during the entire project.

Editor

Advanced Techniques for Microwave Dielectric Characterization

Microwave Open Resonator Techniques – Part I: Theory

Giuseppe Di Massa

Additional information is available at the end of the chapter

1. Introduction

This chapter is devoted to the theory of open resonators. It is well known that lasers uses open resonators as an oscillatory system. In the simplest case, this consists of two mirrors facing each other. This is the first but not the only application of open resonators, whose salient features consist in the fact that their dimensions are much larger than the wavelength and the spectrum of their eigenvalues is much less dense than that of close cavity.

The origin of Open Resonators can be dated to the beginning of the twentieth century when two French physicists developed the classical Fabry-Perot interferometer or Etalon [1]. This novel form of interference device was based on multiple reflection of waves between two closely spaced and highly reflecting mirrors.

In [2] and [3] a theory was developed for resonators with spherical mirrors and approximated the modes by wave beams. The concept of electromagnetic wave beams was also introduced in [5, 12] where was investigated the sequence of lens for the guided transmission of electromagnetic waves.

The use of open resonators either in the microwave region, or at higher frequencies (optical region) has taken place over a number of decades. The related theory and its applications have found a widespread use in several branches of optical physics and today is incorporated in many scientific instruments [6].

In microwave region open resonators have also been proposed as cavities for quasi-optical gyrotrons[16] and as an open cavity in a plasma beat wave accelerator experiment [9].

The use of Open Resonators as microwave Gaussian Beam Antennas [10, 11, 18] provides a very interesting solution as they can provide very low sidelobes level. They are based on the result that the field map at the mid section of an open resonator shows a gaussian distribution that can be used to illuminate a metallic grid or a dielectric sheet.

For microwave applications a reliable description of the coupling between the cavity and the feeding waveguide is necessary. Several papers deal with the coupling through a small hole or a rectangular waveguide taking into account only for the fundamental cavity and waveguide mode [7, 8, 17].

In [4] a complete analysis of the coupling between a rectangular waveguide and an open cavity has been developed taking into account for all the relevant eigenfunctions in the waveguide and in the cavity.

In this paper we review the general theory of Open Resonators and propose to study the coupling between a feeding coupling aperture given by a rectangular or circular waveguide.

Starting from the paraxial approximation of the wave equation, we derive the modal expansion of the field into the cavity taking into account for the proper coordinate system. The computation of the modal coefficients takes into account for the characteristics of the mirrors, the ohmic and diffraction losses and coupling.

2. Open resonator theory

2.1. Parabolic approximation to wave equation

A parabolic equation was first introduced into the analysis of electromagnetic wave propagation in [13] and [14]. Since then, it has been used in diffraction theory to obtain approximate (asymptotic) solutions when the wavelength is small compared to all characteristic dimensions. As open resonators usually satisfy this condition, the parabolic equation finds wide application in developing a theory of open resonators.

A rectangular field component of a coherent wave satisfies the scalar wave equation:

$$\nabla^2 u + k^2 u = 0 \tag{1}$$

where $k = 2\pi/\lambda$ is the propagation constant in the medium.

For a wave traveling in the zeta forward direction, assuming an $e^{j\omega t}$ time dependance, we put:

$$u = \psi(x,y,z)e^{-jkz} \tag{2}$$

where ψ is a slowly varying function which represents the deviation from a plane wave. By inserting (2) into (1) and assuming that ψ varies so slowly with z that its second derivative can be neglected with respect to $\left|k\frac{\partial\psi}{\partial z}\right|$, one obtains the well know parabolic approximation to the wave equation:

$$\frac{\partial^2\psi}{\partial x^2} + \frac{\partial^2\psi}{\partial y^2} - 2jk\frac{\partial\psi}{\partial z} = 0 \tag{3}$$

The differential equation (3), similar to the Schrodinger equation, has solution of the type:

$$\psi = e^{-j\left(P+\frac{k}{2q}r^2\right)} \tag{4}$$

where:

$$r^2 = x^2 + y^2 \tag{5}$$

The parameter $P(z)$ represents a complex phase shift associated to the propagation of the beam along the z axis, $q(z)$ is the complex parameter which describe the Gaussian beam intensity with the distance r from the z axis.

The insertion of (4) in (3) gives the relations:

$$\frac{dq}{dz} = 1 \tag{6}$$

$$\frac{dP}{dz} = -\frac{j}{q} \tag{7}$$

The integration of (6) yields:

$$q(z_2) = q(z_1) + z \tag{8}$$

which relates the intensity in the plane z_2 with the intensity in the plane z_1.

A wave with a Gaussian intensity profile, as (4), is one the most important solutions of equation (3) and is often called *fundamental mode*.

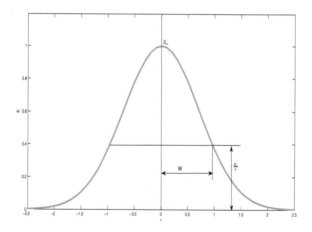

Figure 1. Amplitude distribution of cavity fundamental mode

Two real beam parameters, R and w, are introduced in relation to the above complex parameters q by

$$\frac{1}{q} = \frac{1}{R} - j\frac{\lambda}{\pi w^2} \tag{9}$$

Introducing (9) in the solution (4), we obtain:

$$\psi = e^{-j\left(P + \frac{\pi}{\lambda}\frac{r^2}{R}\right)} e^{-\frac{r^2}{w^2}} \tag{10}$$

Now the physical meaning of these two parameters becomes clear:

- R(z) is the curvature radius of the wavefront that intersects the axis at z;
- w(z) is the decrease of the field amplitude with the distance r from the axis.

Parameter w is called *beam radius* and the term 2w *beam diameter*. The Gaussian beam contracts to a minimum diameter $2w_0$ at beam waist where the phase is plane. The beam parameter q at waist is given by:

$$q_0 = j\frac{\pi w_0}{\lambda} \tag{11}$$

and, using (8), at distance z from the waist:

$$q = q_0 + z = j\frac{\pi w_0}{\lambda} + z \tag{12}$$

Combining (12) and (10), we have:

$$R(z) = z\left[1 + \left(\frac{z_R}{z}\right)^2\right] \tag{13}$$

and

$$w^2(z) = w_0^2\left[1 + \left(\frac{\lambda z}{\pi w_0^2}\right)^2\right] \tag{14}$$

where z_R is the Rayleigh distance:

$$z_R = \frac{\pi w_0^2}{\lambda} \tag{15}$$

The beam contour is an hyperbola with asymptotes inclined to the axis at an angle:

$$\theta = \frac{\lambda}{\pi w_0} \tag{16}$$

In (14) w is the beam radius, w_0 is the minimum beam radius (called beam waist) where one has a plane phase front at $z = 0$ and R is the curvature radius of of the phase front at z. It should be noted that the phase front is not exactly spherical; therefore, its curvature radius is exactly equal to R only on the z-axis. The parameter of the Gaussian beam are illustrated in Fig. 2.

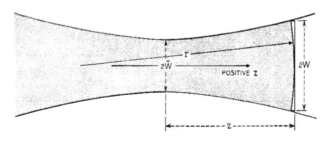

Figure 2. Parameters of Gaussian beam

Dividing (14) by (13), the useful relation is obtained:

$$\frac{\lambda}{\pi w_0^2} = \frac{\pi w^2}{\lambda R} \tag{17}$$

The expression (17) is used to express w_0 and z in terms of w and R:

$$w_0^2 = \frac{w^2}{1 + \left(\frac{\pi w^2}{\lambda R}\right)^2} \tag{18}$$

$$z = \frac{R}{1 + \left(\frac{\lambda R}{\pi w^2}\right)^2} \tag{19}$$

Inserting (11) in (7) we obtain the complex phase shift at distance z from the waist:

$$\frac{dP}{dz} = -\frac{j}{z + j\frac{\pi w_0^2}{\lambda}} \tag{20}$$

Integration of (20) yields

$$jP(z) = \lg\left[1 - j\left(\frac{\lambda z}{\pi w_0^2}\right)\right] = \lg\sqrt{1 + \left(\frac{\lambda z}{\pi w_0^2}\right)^2} - j\arctan\left(\frac{\lambda z}{\pi w_0^2}\right) \tag{21}$$

The real part of P represent the phase shift difference Φ between the Gaussian beam and an ideal plane wave, while the imaginary part produces an amplitude factor $\frac{w_0}{w}$ which gives the decrease of intensity due to the expansion of the beam. Now we can write the fundamental Gaussian beam:

$$u(r,z) = \frac{w_0}{w}e^{\left\{-j(kz-\Phi)-r^2\left(\frac{1}{w^2}+\frac{jk}{2R}\right)\right\}} \tag{22}$$

where:

$$\Phi = \arctan\left(\frac{\lambda z}{\pi w_0^2}\right) \tag{23}$$

2.2. Stability of open resonator

A resonator with spherical mirrors of unequal curvature is representable as a periodic sequence of lens which can be stable or unstable. The stability condition assumes the form:

$$0 < \left(1 - \frac{2l}{R_1}\right)\left(1 - \frac{2l}{R_2}\right) < 1 \tag{24}$$

The above expression was previously derived in [3] from geometrical optics approach based on equivalence of the resonator and a periodic sequence of parallel lens and independently in [5] solving the integral equation for the field distribution of the resonant modes in the limit of infinite Fresnel numbers.

To show graphically which type of resonator is stable and which is unstable, it is useful to plot a stability diagram on which each type of resonator type is represented by a point (Fig. 3).

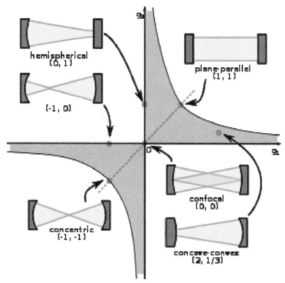

Figure 3. Stability diagram. $g_1 = 1 - \frac{2l}{R_1}$, $g_2 = 1 - \frac{2l}{R_2}$.

2.3. Spherical cavity in cartesian coordinates

In cartesian (x, y, z) coordinates, the separate solutions for (3) are [12]:

$$\psi_{mp}(x,y,z) = \phi_{mp}(x,y,z)exp\left[j(m+p+1)\tan^{-1}\frac{z}{z_R} - j\frac{\pi}{\lambda}\frac{r^2}{R(z)}\right] \quad (25)$$

where[1]

$$\phi_{mp}(x,y,z) = \frac{1}{w(z)}\sqrt{\frac{2}{\pi 2^{m+p}m!p!}}H_m\left(\sqrt{2}\frac{x}{w}\right)H_p\left(\sqrt{2}\frac{y}{w}\right)exp\left[-\frac{r^2}{w^2(z)}\right] \quad (26)$$

H_m is a Hermite polynomial of order m (Appendix A).

Note that both the ϕ_{mp} and ψ_{mp} functions are orthonormal on the transverse planes z=cost.

When we assume that the mirrors are sufficiently large to permit the total reflection of the gaussian beams of any relevant order, we can put:

$$\Psi_{mpq} = u_{mpq}^{(+)} + u_{mpq}^{(-)} \quad (27)$$

---

[1] To be consistent with the parabolic approximation the condition $|m + n + 1| << (kw_0)^2$ must be satisfied.

where $u_{mpq}^{(+)}$ and $u_{mpq}^{(-)}$ represent Hermite Gauss beams propagating from left to right and from right to left,respectively.

Resonance occurs when the phase shift from one mirror to the other is a multiple of π. Using (2), (4) and (9) this condition leads to:

$$k_{mpq}2l - 2(m + p + 1)\tan^{-1}\left(\frac{l}{z_R}\right) = \pi(q + 1) \tag{28}$$

where q is the number of nodes of the axial standing wave pattern and $2l \gg z_R$ the distance between the mirrors (Fig.4).

The fundamental beat frequency Δf_0, i.e. the frequency spacing between successive longitudinal resonances, is given by:

$$\Delta f_0 = \frac{c}{4l} \tag{29}$$

where c is the velocity of light. From (10) the resonant frequency f of a mode can be expressed as:

$$\frac{f_{mpq}}{\Delta f_0} = q + 1 + \frac{1}{\pi}(m + p + 1)\cos^{-1}\left(1 - \frac{2l}{R}\right) \tag{30}$$

The combined use of eqs. (2),(4) and (9) yields:

$$\Psi_{mpq}(x, y, z) = \phi_{mp}(x, y, z)\cos\left[k_{mpq}z - (m + s + 1)\tan^{-1}\frac{z}{z_R} + \frac{\pi}{\lambda}\frac{r^2}{R(z)} + \frac{q\pi}{2}\right] \tag{31}$$

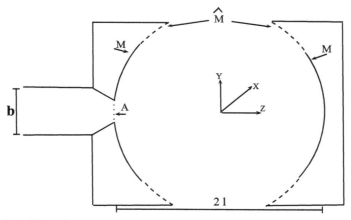

Figure 4. Spherical Open Cavity

In the paraxial approximation the eigenfunctions Ψ_{spq} satisfy the normalization relation

$$\iiint\limits_{cavity} \Psi_{mpq}\Psi_{nst}^* dxdydz = \begin{cases} 1, & mpq \equiv nst; \\ 0, & mpq \neq nst. \end{cases} \tag{32}$$

3. Coupling to feeding waveguide

Because the expressions for the solution of scalar wave equation (3) with the boundary condition $\Phi = 0$ on the mirrors is given by (31) the electromagnetic field inside the cavity can be expressed in terms of (quasi) transverse electromagnetic modes (TEM):

$$\mathbf{E} = \sum_n V_n \mathbf{e}_n \tag{33}$$

$$\mathbf{H} = \sum_n I_n \mathbf{h}_n \tag{34}$$

According to the results reported in sect. 2, the expressions for the $\hat{\mathbf{y}}$ polarized modes are:

$$\mathbf{e}_n = \phi_{mp}(x,y,z) \cos\left[k_{mpq}z - (m+p+1)\tan^{-1}\frac{z}{z_R} + \frac{\pi}{\lambda}\frac{r^2}{R(z)} + \frac{q\pi}{2}\right]\hat{\mathbf{y}}$$

$$\mathbf{h}_n = -\phi_{mp}(x,y,z) \sin\left[k_{mpq}z - (m+p+1)\tan^{-1}\frac{z}{z_R} + \frac{\pi}{\lambda}\frac{r^2}{R(z)} + \frac{q\pi}{2}\right]\hat{\mathbf{x}}$$

and similarly for the $\hat{\mathbf{x}}$ polarized ones.

In (33) the index n summarizes the indexes (mpq).

From Maxwell equations we get for the mode vectors [15]:

$$k_n \mathbf{h}_n = \nabla \times \mathbf{e}_n \tag{35}$$

and for the coefficients V_n, I_n

$$I_n = \frac{j\omega\varepsilon_0}{k^2 - k_n^2}\frac{1}{l}\iint_S \hat{\mathbf{n}} \times \mathbf{E} \cdot \mathbf{h}_n^* dS \tag{36}$$

$$V_n = \frac{k_n}{k^2 - k_n^2}\frac{1}{l}\iint_S \hat{\mathbf{n}} \times \mathbf{E} \cdot \mathbf{h}_n^* dS = -j\frac{\zeta_0\omega_n}{\omega}I_n \tag{37}$$

where S is the cavity surface, $*$ denotes the complex conjugate and ζ_0 is the free space impedance. Note explicitly that the tangential electric field appearing in expression (36) and (37) is the actual field over S. This is not given by expression (33), which, at variance with expression (34), does not provide a representation for the tangential components uniformly valid up to the cavity boundaries. Let us divide the surface S into three parts: the coupling aperture A, the mirrors M and the (ideal) cavity boundary external to the the mirrors, $\hat{M}$ (see Fig. 4). Hence:

$$I_n = \frac{j\omega\varepsilon_0}{k^2 - k_n^2}\frac{1}{l} \cdot \left\{\iint_A \hat{\mathbf{n}} \times \mathbf{E} \cdot \mathbf{h}_n^* dS + \iint_M \hat{\mathbf{n}} \times \mathbf{E} \cdot \mathbf{h}_n^* dS + \iint_{\hat{M}} \hat{\mathbf{n}} \times \mathbf{E} \cdot \mathbf{h}_n^* dS\right\} \tag{38}$$

Strictly speaking, in one of the integrals over the mirrors surfaces the coupling aperture should be deleted. However, the error that we make in extending the integration to the whole mirror is negligible provided that the waveguide dimension is much smaller than that of the mirrors.

The Leontovič boundary condition:

$$\hat{\mathbf{n}} \times \mathbf{E} = \frac{1+j}{\sigma\delta} \hat{\mathbf{n}} \times \mathbf{H} \times \hat{\mathbf{n}} \tag{39}$$

wherein σ is the electric conductivity of the mirrors and $\delta = \sqrt{\frac{2}{\omega\sigma\mu}}$ is the penetration depth, can be applied to express the (tangential) electric field over the mirrors in terms of magnetic one, given by (16). Hence:

$$\iint_M \hat{\mathbf{n}} \times \mathbf{E} \cdot \mathbf{h}_n^* dS = \sum_m I_m \frac{1+j}{\sigma\delta} \iint_M \mathbf{h}_m \cdot \mathbf{h}_n^* dS = 2\frac{1+j}{\sigma\delta} \sum_m \alpha_{nm} I_m \tag{40}$$

being:

$$\alpha_{nm} = \frac{1}{2} \iint_M \mathbf{h}_n^* \cdot \mathbf{h}_m dS \tag{41}$$

Outside the mirrors we can assume that the field is an outgoing locally plane wave (Fresnel-Kirchhoff approximation), so that on $\hat{M}$:

$$\hat{\mathbf{n}} \times \mathbf{E} = \zeta_0 \hat{\mathbf{n}} \times \mathbf{H} \times \hat{\mathbf{n}} \tag{42}$$

Accordingly, we have for the last integral in (38):

$$\iint_{\hat{M}} \hat{\mathbf{n}} \times \mathbf{E} \cdot \mathbf{h}_n^* dS = \zeta_0 \sum_m I_m \iint_{\hat{M}} \mathbf{h}_m \cdot \mathbf{h}_n^* dS = 2\zeta_0 \sum_m I_m (\delta_{nm} - \alpha_{nm}) \tag{43}$$

The last equality in (43) follows from the fact that:

$$\iint_{M \cup \hat{M}} \mathbf{h}_m \cdot \mathbf{h}_n^* dS \simeq \iint_{z=-l} \mathbf{h}_m \cdot \mathbf{h}_n^* dx dy + \iint_{z=l} \mathbf{h}_m \cdot \mathbf{h}_n^* dx dy = 2\delta_{nm} \tag{44}$$

according to the orthogonality condition satisfied by the ϕ functions.

The cavity quality factor Q_n for the n-th mode is defined as:

$$Q_n = \omega_n \frac{W_n}{P_n} \tag{45}$$

where W_n is the mean electromagnetic energy stored in the cavity and P_n is the power lost when only the nth mode is excited at the resonance pulsation ω_n. The power is lost due to diffraction and ohmic losses.

By taking (32) into account, we can express the diagonal term in (40) as a function of the quality factor for the ohmic losses, Q_{rn}:

$$2\alpha_{nn} = \iint_M |\mathbf{h}_n|^2 dS = \frac{\omega_n \mu_0}{Q_{rn}} \sigma \delta_n l \tag{46}$$

being δ_n the skin depth at the resonant frequency.

The diffraction losses of a cavity can be calculated by taking into account for the diffraction effects produced by the finite size of the mirrors. Under the simplifying assumption of quasi-optic nature of the problem (dimensions of the resonator large compared to wavelength and quasi-transverse electromagnetic fields) the Fresnel-Kirchhoff formulation can be invoked for the diffracted field from the mirrors. Hence we have for the diffraction loss of the nth mode:

$$P_{dn} = Re \left(\iint_{\hat{M}} \frac{1}{2} \mathbf{E}_n \times \mathbf{H}_n^* \cdot \hat{n} dS \right) = \frac{1}{2} \zeta_0 |I_n|^2 \iint_{\hat{M}} |\mathbf{h}_n|^2 ds \qquad (47)$$

and for the corresponding quality factor for the diffraction losses, Q_{d}:

$$\frac{\zeta_0}{l} \iint_{\hat{M}} |\mathbf{h}_n|^2 dS = \frac{\omega_n \mu_0}{Q_{dn}} \qquad (48)$$

By using (40, 41,44, 45) and taking into account that $\sigma\delta\zeta_0 >> 1$ and $\delta_n/\delta \simeq 1$ for all relevant modes, equation (38) becomes:

$$I_n = \frac{j\omega\varepsilon_0}{\left(k^2 - k_n^2 + \frac{kk_n}{Q_{rn}} \right) - j\frac{kk_n}{Q_{Tn}}} \frac{1}{l} \left\{ \iint_A \mathbf{E} \times \mathbf{h}_n^* \cdot \hat{n} dS - 2\zeta_0 \sum_m {}' I_m \alpha_{nm} \right\} \qquad (49)$$

wherein $\sum' \equiv \sum_{n \neq m}$ and:

$$\frac{1}{Q_{Tn}} = \left(\frac{1}{Q_{dn}} + \frac{1}{Q_{rn}} \right) \qquad (50)$$

A metallic waveguide is assumed to feed the cavity. The waveguide field on the coupling aperture A is represented as:

$$\mathbf{E}^g = \sum_n V_n^g \mathbf{e}_n^g \qquad (51)$$

$$\mathbf{H}^g = \sum_n I_n^g \mathbf{h}_n^g \qquad (52)$$

where $\mathbf{e}^g$ and $\mathbf{h}^g$ are TE electromagnetic modes of the waveguide:

Assuming that the mirror curvature can be neglected over the extension of the coupling aperture, fields (51, 52) verify the following orthonormality relation:

$$\iint_A \mathbf{e}_n^g \times \mathbf{h}_m^g \cdot \hat{z} dS = \delta_{nm} \qquad (53)$$

Expressing the field over the coupling aperture A by means of expression (52) we obtain from (3):

$$I_n + 2F_n \sum_m {}' \alpha_{nm} I_m = \frac{F_n}{\zeta_0} \sum_m \beta_{nm} V_m^g \qquad (54)$$

where:

$$F_n = \frac{jk/l}{\left(k^2 - k_n^2 + \frac{kk_n}{Q_{rn}} \right) - j\frac{kk_n}{Q_{Tn}}} \qquad (55)$$

and:

$$\beta_{nm} = \iint_A \mathbf{e}_m^g \times \mathbf{h}_n^* \cdot \hat{n} dS = -\iint_A \mathbf{h}_n^* \cdot \mathbf{h}_m^g dS. \qquad (56)$$

By introducing the matrices $\underline{A}$ and $\underline{B}$, whose elements are:

$$a_{nm} = \begin{cases} \frac{1}{F_n} & n=m \\ 2\alpha_{nm} & n \neq m. \end{cases} \qquad (57)$$

and β_{nm} respectively [2], and the vectors $\underline{I} \equiv \{I_n\}$ and $\underline{V}^g \equiv \{V_n^g\}$, relation (3) can be written in a compact form as:

$$\zeta_0 \underline{\underline{A}} \cdot \underline{I} = \underline{\underline{B}} \cdot \underline{V}^g = \underline{\underline{B}} \cdot \left(\underline{V}^+ + \underline{V}^- \right) \tag{58}$$

wherein $\underline{V}^+$ and $\underline{V}^-$ are the vectors of the incident and reflected waveguide mode amplitudes respectively. By enforcing the continuity of the magnetic field tangential component over the coupling aperture, we get:

$$- \underline{\underline{B}}^+ \cdot \underline{I} = \underline{I}^g = \frac{1}{\zeta_0} \underline{\underline{\zeta}}^{-1} \cdot \left(\underline{V}^+ - \underline{V}^- \right) \tag{59}$$

wherein $\underline{\underline{B}}^+$ is the adjoint (i.e., the transpose, being $\underline{\underline{B}}$ a real matrix) of $\underline{\underline{B}}$ and $\underline{\underline{\zeta}}$ is the diagonal matrix whose elements are the modes characteristic impedances, normalized to ζ_0. From (48) and (49) we immediately obtain:

$$\left(\underline{\underline{I}} - \underline{\underline{A}}^{-1} \cdot \underline{\underline{B}} \cdot \underline{\underline{\zeta}} \cdot \underline{\underline{B}}^+ \right) \cdot \underline{I} = \frac{2}{\zeta_0} \underline{\underline{A}}^{-1} \cdot \underline{\underline{B}} \cdot \underline{V}^+ \tag{60}$$

$$\left(\underline{\underline{I}} - \underline{\underline{\zeta}} \cdot \underline{\underline{B}}^+ \cdot \underline{\underline{A}}^{-1} \cdot \underline{\underline{B}} \right) \cdot \underline{V}^- = \left(\underline{\underline{I}} + \underline{\underline{\zeta}} \cdot \underline{\underline{B}}^+ \cdot \underline{\underline{A}}^{-1} \cdot \underline{\underline{B}} \right) \cdot \underline{V}^+ \tag{61}$$

wherein $\underline{\underline{I}}$ is the unit matrix and $\underline{\underline{A}}^{-1}$ the inverse of the matrix $\underline{\underline{A}}$.

Solution of eq. (60) and (61) provides the answer to our problem. In particular, from (61) we get the (formal) expression for the feeding waveguide scattering matrix $\mathcal{S}$:

$$\underline{\underline{S}} = \left(\underline{\underline{I}} - \underline{\underline{\zeta}} \cdot \underline{\underline{B}}^+ \cdot \underline{\underline{A}}^{-1} \cdot \underline{\underline{B}} \right)^{-1} \cdot \left(\underline{\underline{I}} + \underline{\underline{\zeta}} \cdot \underline{\underline{B}}^+ \cdot \underline{\underline{A}}^{-1} \cdot \underline{\underline{B}} \right) \tag{62}$$

4. Field on the coupling aperture

In order to solve the system (60-61) a suitable description of the field on the aperture A is necessary. Any device, able to support electromagnetic field matching the cavity field on the mirror (33-34), can be used to feed the cavity. In the following, the case of metallic and circular waveguide will be treated in detail.

4.1. Modes in rectangular waveguide

A rectangular metallic waveguide, with transverse dimensions $a \times b$, is assumed to feed the cavity. The waveguide TE electromagnetic modes of the metallic rectangular waveguide, on the coupling aperture A, is represented as:

$$\mathbf{h}_n = \mathbf{h}_{pq} = \frac{1}{k_{tpq}} \sqrt{\frac{4 \varepsilon_p \varepsilon_q}{ab}} \cdot \tag{63}$$

$$\left\{ \frac{p\pi}{a} \sin \frac{p\pi}{a} \left(x + \frac{a}{2} \right) \cos \frac{q\pi}{b} \left(y + \frac{b}{2} \right) \hat{\mathbf{x}} + \frac{q\pi}{b} \cos \frac{p\pi}{a} \left(x + \frac{a}{2} \right) \sin \frac{q\pi}{b} \left(y + \frac{b}{2} \right) \hat{\mathbf{y}} \right\}$$

---

[2] Note explicitly that α_{nm} and β_{nm} are real quantities, as both the cavity and waveguide mode vectors are real. Moreover $\alpha_{nm} = \alpha_{mn}$ so that the matrix $\underline{\underline{A}}$ is symmetric.

$$\mathbf{e}_{pq} = \mathbf{h}_{pq} \times \hat{\mathbf{z}} \tag{64}$$

$$\varepsilon_p = \begin{cases} 1, & p \neq 0; \\ \frac{1}{2}, & p = 0. \end{cases} \tag{65}$$

$$k_{tpq}^2 = \left(\frac{p\pi}{a}\right)^2 + \left(\frac{q\pi}{b}\right)^2 \tag{66}$$

and the TM electromagnetic modes:

$$\mathbf{e}_n = \mathbf{e}_{pq} = -\frac{1}{k_{tpq}} \sqrt{\frac{4\varepsilon_p \varepsilon_q}{ab}} \cdot \tag{67}$$

$$\left\{ \frac{p\pi}{a} \cos\frac{p\pi}{a}\left(x+\frac{a}{2}\right) \sin\frac{q\pi}{b}\left(y+\frac{b}{2}\right)\hat{\mathbf{x}} + \frac{q\pi}{b}\sin\frac{p\pi}{a}\left(x+\frac{a}{2}\right)\cos\frac{q\pi}{b}\left(y+\frac{b}{2}\right)\hat{\mathbf{y}} \right\}$$

$$\mathbf{h}_{pq} = \hat{\mathbf{z}} \times \mathbf{e}_{pq} \tag{68}$$

Note explicitly that in expressions (63-68) the index n summarizes the indexes (pq).

4.2. Modes in circular waveguide

When a circular waveguide , with radius a, is assumed to feed the cavity, the TE electromagnetic modes are:

$$\mathbf{h}_n = \mathbf{h}_{pr} = -\sqrt{\frac{\xi_p}{\pi}} \frac{1}{\sqrt{q_{pr}'^2 - p^2}} \frac{1}{J_p(q_{pr}')} \cdot$$

$$\left\{ \left[\frac{q_{pr}'}{a} J_p'(k_{tpr}'\rho)\cos\phi \begin{Bmatrix} \cos(p\phi) \\ \sin(p\phi) \end{Bmatrix} - \frac{p}{\rho}J_p(k_{tpr}'\rho)\sin\phi \begin{Bmatrix} \sin(p\phi) \\ -\cos(p\phi) \end{Bmatrix} \right]\hat{\mathbf{x}} \right.$$

$$\left. + \left[\frac{q_{pr}'}{a} J_p'(k_{tpr}'\rho)\sin\phi \begin{Bmatrix} \cos(p\phi) \\ \sin(p\phi) \end{Bmatrix} + \frac{p}{\rho}J_p(k_{tpr}'\rho)\cos\phi \begin{Bmatrix} \sin(p\phi) \\ -\cos(p\phi) \end{Bmatrix} \right]\hat{\mathbf{y}} \right\} \tag{69}$$

$$\mathbf{e}_{pr} = \mathbf{h}_{pr} \times \hat{\mathbf{z}} \tag{70}$$

and the TM electromagnetic modes:

$$\mathbf{e}_n = \mathbf{e}_{pr} = -\sqrt{\frac{\xi_p}{\pi}} \frac{1}{J_{p+1}'(q_{pr})} \cdot$$

$$\left\{ \left[\frac{1}{a} J_p'(k_{tpr}\rho)\cos\phi \begin{Bmatrix} \cos(p\phi) \\ \sin(p\phi) \end{Bmatrix} + \frac{p}{q_{pr}\rho}J_p(k_{tpr}\rho)\sin\phi \begin{Bmatrix} \sin(p\phi) \\ -\cos(p\phi) \end{Bmatrix} \right]\hat{\mathbf{x}} \right.$$

$$\left. + \left[\frac{1}{a} J_p'(k_{tpr}\rho)\sin\phi \begin{Bmatrix} \cos(p\phi) \\ \sin(p\phi) \end{Bmatrix} - \frac{p}{q_{pr}\rho}J_p(k_{tpr}\rho)\cos\phi \begin{Bmatrix} \sin(p\phi) \\ -\cos(p\phi) \end{Bmatrix} \right]\hat{\mathbf{y}} \right\} \tag{71}$$

$$\mathbf{h}_{pr} = \hat{\mathbf{z}} \times \mathbf{h}_{pr} \tag{72}$$

where:

$$k_{tpr}^2 = \left(\frac{q_{pr}}{a}\right)^2; \quad k_{tpr}'^2 = \left(\frac{q_{pr}'}{a}\right)^2 \tag{73}$$

q_{or} is the r-mo zero of Bessel function of order p and q'_{or} r-mo zero of the derivative of Bessel function of order p.

$$\xi_p = \begin{cases} 1, \ p = 0; \\ 2, \ p \neq 0. \end{cases} \tag{74}$$

Note explicitly that in expressions (69, 71) the index n summarizes the indexes (pr).

5. Equivalent circuit

Let us consider systems (60) and (61) under the following assumptions:

1. Negligible intercoupling between cavity modes, i.e. :

$$(\underline{A})_{pq} = \frac{1}{F_p}\delta_{pq} \Longleftrightarrow (\underline{A}^{-1})_{pq} = F_p\delta_{pq} \tag{75}$$

2. Single cavity mode approximation, i.e., -see (3)-:

$$F_p = \delta_{p0}F_0 \tag{76}$$

3. Beam diameter at the mirror much larger than the waveguide dimension, i.e. :

$$w = w(l) \gg a \tag{77}$$

Putting

$$V_n^g = V_1^+\delta_{1n} + V_n^- \tag{78}$$

and taking (75-76) into account, equations (60-61) became

$$-\zeta_0\zeta_n\beta_{0n}I_0 = V_1^+\delta_{0n} - V_n^- \tag{79}$$

$$\left(1 - F_0 \sum_n \beta_{0n}^2\zeta_n\right)I_0 = \frac{2F_0}{\zeta_0}\beta_{01}V_1^+ \tag{80}$$

From (79,80) we immediately get:

$$V_n^- = \left(\delta_{1n} + \frac{2F_0\zeta_n\frac{\beta_{0n}}{\beta_{01}}}{1 - F_0 \sum_k \beta_{0k}^2\zeta_k}\right)V_1^+ \tag{81}$$

Hence

$$\Gamma = \frac{V_1^-}{V_1^+} = \frac{1 + F_0\zeta_1\beta_{01}^2 - F_0 \sum_{k\neq1}\zeta_k\beta_{0k}^2}{1 - F_0 \sum_k \zeta_k\beta_{0k}^2} \tag{82}$$

From (82) we get for the equivalent terminal impedance relative to the fundamental mode:

$$Z = \zeta_0\zeta_1\frac{1 + \Gamma}{1 - \Gamma} = -\frac{\zeta_0}{\beta_{01}^2 F_0} + \zeta_0 \sum_{k\neq1}\left(\frac{\beta_{0k}}{\beta_{01}}\right)^2\zeta_k \tag{83}$$

Taking into account for the expression (55) for F_0, we get the equivalent circuit representation of Fig. 5.

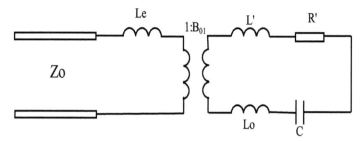

Figure 5. Equivalent Circuit

where the explicit expression for its elements are collected under Table 1.

L_0	$\mu_0 l$	(H)	C	$\varepsilon l / (k_0 l)^2$ (F)
L'	$\mu_0 \delta$	(H)	R'	$2/\sigma\delta$ (Ω)

Table 1. Expressions for the circuit elements of Fig.5

The value of L_e depends on the feeding waveguide and is reported in the following subsections for rectangular and circular waveguides.

5.1. Rectangular waveguide

The cavity is assumed to be excited by the incident fundamental TE_{10} mode. From expressions (51-52) and (63) follows that in the cavity are excited only the (0,0,q) mode, in the feeding waveguide are excited the modes TE_{n0}.

When the cavity is fed by a rectangular waveguide under the approximation 3), the expression (56) for β_{0k} can be explicitly evaluated, leading to:

$$\beta_{0n} = \frac{4}{\pi}\frac{1}{n}\sqrt{\frac{ab}{w^2}} \quad n = 1, 3, \cdots \tag{84}$$

Hence, for the sum at the right hand side of (83) we have:

$$\zeta_0 \sum_{k=3,5,\cdots} \left(\frac{\beta_{0k}}{\beta_{01}}\right)^2 \zeta_k = j\zeta_0 \sum_{k=0}^{\infty} \frac{1}{(2k+3)^2} \frac{1}{\sqrt{\left(\frac{2k+3}{2a}\lambda\right)^2 - 1}} \simeq$$

$$\simeq j\zeta_0 \left(\frac{2a}{\lambda}\right) \sum_{k=0}^{\infty} \frac{1}{(2k+3)^3} = j\omega\mu a \frac{1}{\pi}\left[\left(1 - \frac{1}{8}\right)\zeta(3) - 1\right] = j\omega L_e \tag{85}$$

wherein $\zeta(\cdot)$ denotes the Rieman zeta function. From equation (85), we have the value of L_e:

$$L_e = 16.5 \ 10^{-3}\mu_0 \ a \ [henry] \tag{86}$$

5.2. Circular waveguide

The cavity is assumed to be excited by the incident fundamental TE_{11} mode. From expressions (51, 52) and (69) follows that in the cavity are excited the (0,0,q) mode, in the feeding waveguide are excited the modes TE_{1r}.

When the cavity is fed by a circular waveguide under the approximation 3), the expression (56) for β_{0k} can be evaluated (Appendix A), leading to:

$$\beta_{0r} = -\frac{2}{\pi} \frac{1}{\sqrt{q_{pr}'^2 - p^2}} \frac{1}{J_p(q_{pr}')} \frac{1}{w(l)} \cdot [I_1 - I_2] \tag{87}$$

$$I_1 = \pi \frac{q_{pr}'}{a} \int_0^a J_1'(k_{t1r}'\rho) e^{\left[-\frac{\rho^2}{w^2(l)}\right]} \sin\left[k_{00q}l - \tan^{-1}\frac{l}{z_R} + \frac{\pi}{\lambda}\frac{\rho^2}{R(l)} + \frac{q\pi}{2}\right] \rho\, d\rho \tag{88}$$

$$I_2 = \pi \int_0^a J_1(k_{t1r}'\rho) e^{\left[-\frac{\rho^2}{w^2(l)}\right]} \sin\left[k_{00q}l - \tan^{-1}\frac{l}{z_R} + \frac{\pi}{\lambda}\frac{\rho^2}{R(l)} + \frac{q\pi}{2}\right] d\rho \tag{89}$$

allowing the computation of L_e.

Author details

Giuseppe Di Massa
University of Calabria, Italy

Appendix

A. Hermite polynomials

The Hermite polynomials are defined as:

$$H_n(x) = (-1)^n e^{x^2} \frac{\partial^n}{\partial x^n} e^{-x^2} \tag{90}$$

The differential equation:

$$\frac{\partial^2 y}{\partial x^2} - 2x\frac{\partial y}{\partial x} + 2y = 0 \tag{91}$$

admits as solution the Hermite polynomial $H_n(x)$. For the the Hermite polynomials the following orthogonal relation holds:

$$\int_{-\infty}^{+\infty} e^{-\frac{x^2}{2}} H_n(x) H_m(x) = \begin{cases} \sqrt{\pi}2^n n!, & \text{n=m;} \\ 0, & \text{n} \neq \text{m.} \end{cases} \tag{92}$$

In the following some particular values with the recursion relation are reported:

$$H_0(x) = 1$$
$$H_1(x) = 2x$$
$$H_2(x) = 4x^2 - 2$$

$$H_3(x) = 8x^3 - 12x$$
$$H_4(x) = 16x^4 - 48x^2 + 12$$
$$- - - \quad - - - - - - -$$
$$H_{n+1} = 2xH_n(x) - 2nH_{n-1}(x)$$

B. Coupling cavity - circular waveguide

The x component of the magnetic field, for TE modes, in the waveguide is:

$$\mathbf{h}^g_{\mathbf{x}pr} = -\sqrt{\frac{\tilde{\zeta}p}{\pi}} \frac{1}{\sqrt{q'^2_{pr} - p^2}} \frac{1}{J_p(q'_{pr})} \cdot \tag{93}$$

$$\left[\frac{q'_{pr}}{a} J'_p(k'_{tpr}\rho) \cos\phi \cos(p\phi) - \frac{p}{\rho} J_p(k'_{tpr}\rho) \sin\phi \sin(p\phi) \right]$$

The x component of the magnetic field in the cavity is:

$$\mathbf{h}_{\mathbf{x}n} = -\phi_{mp}(x,y,z) \sin\left[k_{mpq}z - (m+p+1)\tan^{-1}\frac{z}{z_R} + \frac{\pi}{\lambda}\frac{r^2}{R(z)} + \frac{q\pi}{2} \right] \tag{94}$$

where

$$\phi_{mp}(x,y,z) = \frac{1}{w(z)}\sqrt{\frac{2}{\pi 2^{m+p}m!p!}} H_m\left(\sqrt{2}\frac{x}{w}\right) H_p\left(\sqrt{2}\frac{y}{w}\right) exp\left[-\frac{r^2}{w^2(z)}\right] \tag{95}$$

According to the position of sect.4 we consider the mode (0,0,q) in the cavity, so the equation (95) reduces to:

$$\phi_{00}(x,y,z) = \frac{1}{w(z)}\sqrt{\frac{2}{\pi}} exp\left[-\frac{r^2}{w^2(z)}\right] \tag{96}$$

and the equation (94) on the mirror (for z=l):

$$\mathbf{h}_{\mathbf{x}n} = -\frac{1}{w(l)}\sqrt{\frac{2}{\pi}}e^{\left[-\frac{r^2}{w^2(l)}\right]} \sin\left[k_{00q}l - \tan^{-1}\frac{l}{z_R} + \frac{\pi}{\lambda}\frac{r^2}{R(l)} + \frac{q\pi}{2} \right] \tag{97}$$

$$\beta_{0m} = -\iint_A \mathbf{h}^*_0 \cdot \mathbf{h}^g_m dS \tag{98}$$

$$= -\frac{2}{\pi}\frac{1}{\sqrt{q'^2_{pr} - p^2}}\frac{1}{J_p(q'_{pr})}\frac{1}{w(l)} \cdot \tag{99}$$

$$\iint_A \left[\frac{q'_{pr}}{a} J'_p(k'_{tpr}\rho) \cos\phi \cos(p\phi) - \frac{p}{\rho} J_p(k'_{tpr}\rho) \sin\phi \sin(p\phi) \right] \cdot \tag{100}$$

$$e^{\left[-\frac{\rho^2}{w^2(l)}\right]} \sin\left[k_{00q}l - \tan^{-1}\frac{l}{z_R} + \frac{\pi}{\lambda}\frac{\rho^2}{R(l)} + \frac{q\pi}{2} \right] \rho d\rho d\phi \tag{101}$$

$$\beta_{0r} = -\frac{2}{\pi}\frac{1}{\sqrt{q_{pr}'^2 - p^2}}\frac{1}{J_p(q_{pr}')}\frac{1}{w(l)} \cdot [I_1 - I_2] \tag{102}$$

where:

$$I_1 = p\frac{\sin(2p\pi)}{p^2 - 1}\frac{q_{pr}'}{a}\int_0^a J_p'(k_{tpr}'\rho)e^{\left[-\frac{\rho^2}{w^2(l)}\right]}\sin\left[k_{00q}l - \tan^{-1}\frac{l}{z_R} + \frac{\pi}{\lambda}\frac{\rho^2}{R(l)} + \frac{q\pi}{2}\right]\rho d\rho \tag{103}$$

$$I_2 = \frac{\sin(2p\pi)}{p^2 - 1}\int_0^a \frac{p}{\rho}J_p(k_{tpr}'\rho)e^{\left[-\frac{\rho^2}{w^2(l)}\right]}\sin\left[k_{00q}l - \tan^{-1}\frac{l}{z_R} + \frac{\pi}{\lambda}\frac{\rho^2}{R(l)} + \frac{q\pi}{2}\right]\rho d\rho \tag{104}$$

that are not equal to zero for p=1, giving:

$$I_1 = \pi\frac{q_{pr}'}{a}\int_0^a J_1'(k_{t1r}'\rho)e^{\left[-\frac{\rho^2}{w^2(l)}\right]}\sin\left[k_{00q}l - \tan^{-1}\frac{l}{z_R} + \frac{\pi}{\lambda}\frac{\rho^2}{R(l)} + \frac{q\pi}{2}\right]\rho d\rho \tag{105}$$

$$I_2 = \pi\int_0^a J_1(k_{t1r}'\rho)e^{\left[-\frac{\rho^2}{w^2(l)}\right]}\sin\left[k_{00q}l - \tan^{-1}\frac{l}{z_R} + \frac{\pi}{\lambda}\frac{\rho^2}{R(l)} + \frac{q\pi}{2}\right]d\rho \tag{106}$$

6. References

[1] C. Fabry and A. Perot, (1899), Theorie et Applications d'une Nouvélle Method de Spectroscopie Interférentielle, Ann Chim. Phys., Vol. 7, Vol. 16, 115-143.

[2] G. D. Boyd and J. P. Gordon, (1961), Confocal multimode resonator for millimeter through outical wavelength masers, Bell Sys. Tech. J.,Vol. 40, 489-508.

[3] G. D. Boyd and H. Kogelnik, (1962), Generalized confocal resonator theory, Bell Sys. Tech. J., Vol. 41, 1347-1369.

[4] O. Bucci, G. Di Massa, (1992), Open resonator powered by rectangular waveguide, IEE Proceedings-H, Vol. 139, 323-329.

[5] P. O. Clark, (1964). A self consistent field analisys of Spherical-Mirror Fabry Perot Resonators, Proc. of the IEEE, Vol. 53, No. 1, 36 - 41.

[6] R. N. Clarke, C. B. Rosemberg, (1982). Fabry-Perot and Open Resonators at Microwave and Millimeter wave frequencies, 2-200 GHz, J. Phys. E: Sci. Instrum.,Vol. 15, 9-24.

[7] A. L. Cullen, P.R. Yu, (1979), Complex source-point theory of the electromagnetic open resonator, Proc. R. Soc Lond. A., Vol. 366, 155-171.

[8] G. Di Massa, D. Cuomo, A. Cutolo, G. Delle Cave (1989), Open resonator for microwave application, IEE Proc. H,Vol. 136, 159-164.

[9] G. Di Massa, R. Fedele, G. Miano, 1990, C. Nappi (1990), A beat wave experiment in open resonator, Physica Scripta, Vol. T30, 122-126.

[10] G. Di Massa,L. Boccia,G. Amendola, (2005), A Gaussian beam antenna based on an open resonator, The European Conference on Wireless Technology, Paris, France.

[11] G. Di Massa,L. Boccia, G. Amendola, (2005),Gaussian beam antennas based on open resonator structures, 28th ESA Antenna Workshop on Space Antenna Systems and Technologies.

[12] H. Kogelnik, T. Li, (1966), Laser beam and resonator, Proc. IEEE, Vol. 54, 1312-1329.

[13] Leontovich M.A.(1944), Statistical Physics, OGIZ, Moskow.

[14] Leontovich M.A., Fock V. A. (1948), Investigation of Radio Wave Propagation, ed. B.A. Vendeski, AN SSSR, Moskow, 13-39.

[15] K. Kurokawa, (1969), An introduction to theory of of microwave circuits, Academic Press, New York.

[16] A. Perrenoud, M.Q. Tran, B. Isac, (1986), On the design of open resonator for quasi-optical gyratron, Int. J. of Infrared and Millimeter Waves, Vol. 7, 427-446.

[17] V. N. Rodionova, A. Ya. Slepyan, G. Ya. Slepian, (1991) Oliner model for quasioptical resonator to rectangular waveguide coupling elements, Electronic Letters, Vol. 27, 1427-1428.

[18] R. Sauleau, P. Coquet, D. Thouroude, J.P. Daniel, and T. Matsui, (2003), Radiation Characteristics and Performance of Millimeter-Wave Horn-Fed Gaussian Beam Antennas, IEEE Trans. Antennas and Propagation, Vol. 51, 379-386.

Chapter 2

Microwave Open Resonator Techniques - Part II: Applications

Sandra Costanzo, Giuseppe Di Massa and Hugo Oswaldo Moreno

Additional information is available at the end of the chapter

1. Introduction

The accurate modelling of dielectric and impedance features [1, 2] is an essential need for the efficient design of microstrip circuits and antennas. The increasing demand for miniaturization and high-frequency operation strongly imposes the accurate characterization of low-loss thin dielectric surfaces [12, 14], both in the form of simple laminated substrates [6, 9, 19, 23], as well as in more complex configurations of microstrip grids to be used for reflectarray/transmitarray structures [3, 21]. Various microwave techniques have been introduced in literature to characterize the electrical properties of materials. They include open-ended waveguide/coaxial probe methods [18], free-space techniques [16], stripline [20], transmission/reflection [4] and resonant [22] procedures, all having specific advantages and constraints. Among them, open resonator methods [10, 11, 13, 15, 17, 25] give the most powerful tool to accurately retrieve the equivalent impedance properties of low-loss thin dielectric surfaces. In the standard resonator approach [25], approximate empirical formulas are adopted to obtain the surface impedance characterization from the knowledge of measured resonance parameters, such as the frequency shift and the cavity quality factor. Information from different sample thicknesses and/or positioning are generally adopted to increase the accuracy in the surface parameters extraction, which is performed in terms of a transcendental equation having multiple roots [11].

All existing papers on open resonator methods inherit the approach proposed in [25], thus assuming an ideal open cavity and neglecting the excitation of higher-order modes, which have a great relevance for an accurate modelling of the coupling with the feeding waveguide. In this chapter, an equivalent circuit formulation is adopted to accurately model the open resonator behavior in the presence of the test surface. On the basis of a complete modal analysis as outlined in [5], the adopted circuit approach leads to optimize the coupling between the cavity and the exciting waveguide, further including a proper modelling of the cavity losses, which provides a significant improvement when compared to the traditional open resonator approach [25].

The chapter is organized as follows. In Section 2, a detailed description of the equivalent circuit formulation is provided and an accurate expression is derived for the equivalent impedance of the test planar surface. In Section 3, the adoption of the proposed approach for the dielectric characterization of thin grounded dielectric substrate is presented. A further application of the method to the phase response characterization of microstrip reflectarray elements is provided in Section 4. Conclusions are finally outlined in Section 5.

2. Equivalent circuit model of open resonator system for the characterization of planar surfaces

The open resonator system adopted for the equivalent impedance characterization is illustrated in Fig. 1. It consists of a spherical mirror of radius R_o, which is used to produce a Gaussian beam impinging on the grounded test surface at a distance l. A feeding rectangular waveguide of standard height b is adopted for the cavity excitation, and a transition giving a final height b_1 is properly designed to optimize the coupling with the cavity.

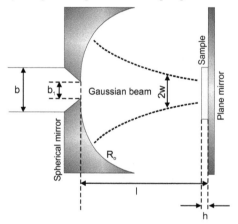

Figure 1. Hemispherical open resonator system

At this purpose, the approach proposed in [5] is adopted, which is based on a complete eigenfunctions analysis leading to the equivalent circuit of Fig. 2 [8].

The relative circuit parameters, accurately defined in [5] and reported in Table 1, properly take into account for the ohmic and the diffraction losses of the cavity. In particular, the component R' models the losses due to the finite conductivity of the mirrors, while the term L' gives the effect of the skin depth δ. It sums to the inductive circuit part L_o modelling the cavity, thus producing a shift in the resonant frequency, which is equivalent to a cavity enlargement.

In Table 1, the term $2l$ gives the cavity length, a is the major waveguide dimension, σ represents the conductivity, while $k_o = 2\pi f_o/c$ is the free-space propagation constant, c being the velocity of light and f_o the resonant frequency of the empty resonator.

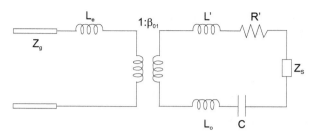

Figure 2. Equivalent circuit of open resonator system

L_o [H]	L' [H]	L_e [H]	C [F]	R' [Ω]
$\mu_o l$	$\mu_o \delta$	$16.5 \cdot 10^{-3} \mu_o a$	$\frac{\epsilon l}{(k_o l)^2}$	$\frac{2}{\sigma \delta}$

Table 1. Expressions of circuit parameters in Fig. 2

From the circuit reported in Fig. 2 it is easy to derive the expressions of the waveguide input impedance Z_i and the reflection coefficient Γ, respectively given as [8]:

$$Z_i = j\omega L_e + \frac{Z_R}{\beta_{01}^2} \tag{1}$$

$$\Gamma = \frac{Z_i - Z_g}{Zi + Z_g} \tag{2}$$

where:

$$Z_R = j\omega L_T + R' + \frac{1}{j\omega C} \tag{3}$$

$$L_T = L_o + L' \tag{4}$$

while β_{01} is the waveguide-cavity factor as reported in [5] and Z_g represents the characteristic impedance of the transmission line equivalent to the rectangular waveguide excited in its fundamental mode TE_{10}.

The insertion of the grounded test surface results into an equivalent impedance Z_S (Fig. 2), which can be easily expressed as [8]:

$$Z_S = jZ_d tan\,(k_d h - \phi_G) \tag{5}$$

The terms Z_d and k_d into eq. (5) respectively give the characteristic impedance and the propagation constant of the shorted transmission line equivalent to the grounded dielectric slab, while the phase shift ϕ_G takes into account for the Gaussian nature of the beam, and is given as [7]:

$$\phi_G = arctan\left(\frac{h}{d_R}\right) \tag{6}$$

where $d_R = \sqrt{Rl - l^2}$ is the Rayleigh distance.

Due to the negligible contribution of the Gaussian amplitude variation, a uniform wave is assumed, leading to use an equivalent circuit approach, while the phase term ϕ_G introduced by the Gaussian beam is properly considered to model the associated resonance frequency shift.

It is straightforward to deduce the equivalent impedance Z_S of eq. (5) from return loss measurements performed at the waveguide input, under empty and loaded cavity conditions. As a matter of fact, the insertion of the test surface sample produces a shift in the resonant frequency of the cavity, from which the imaginary part of impedance Z_S can be derived. At the same time, an amplitude reduction of the reflection coefficient is obtained in the presence of the grounded slab, which in turns is related to real part of impedance Z_S.

In the following Sections, a more detailed description of the impedance reconstruction method is provided for two specific application contexts, namely the complex permittivity retrieval of thin dielectric substrates and the phase response characterization of microstrip reflectarrays.

3. Open resonator application to dielectric material characterization

Let us consider as test surface a grounded dielectric sheet having thickness h (Fig. 1) and unknown complex permittivity $\epsilon = \epsilon' - j\epsilon''$. Under this hypothesis, the equivalent impedance Z_S can be easily expressed as [8]:

$$Z_S = j\frac{Z_0}{a_\epsilon (1 - jb_\epsilon)} tan \{k_0 [a_\epsilon (1 - jb_\epsilon)] h - \phi_G\} \tag{7}$$

where $a_\epsilon = \sqrt{\epsilon'}$, $b_\epsilon = \frac{1}{2}tan\delta$, Z_0 and k_0 being, respectively, the free-space impedance and propagation constant.

After some manipulations, the real and imaginary parts of impedance Z_S can be derived as follows [8]:

$$Re\{Z_S\} = \frac{Z_0}{a_\epsilon (b_\epsilon^2 + 1)} \cdot \frac{sinh(2a_\epsilon b_\epsilon hk_0) - b_\epsilon sin\left\{2\left[a_\epsilon hk_0 + arctan\left(\frac{\lambda h}{\pi w^2}\right)\right]\right\}}{cosh(2a_\epsilon b_\epsilon hk_0) + cos\left\{2\left[a_\epsilon hk_0 + arctan\left(\frac{\lambda h}{\pi w^2}\right)\right]\right\}} \tag{8}$$

$$Im\{Z_S\} = \frac{Z_0}{a_\epsilon (b_\epsilon^2 + 1)} \cdot \frac{sin\left\{2\left[a_\epsilon hk_0 + arctan\left(\frac{\lambda h}{\pi w^2}\right)\right]\right\} + b_\epsilon sinh(2a_\epsilon b_\epsilon hk_0)}{cosh(2a_\epsilon b_\epsilon hk_0) + cos\left\{2\left[a_\epsilon hk_0 + arctan\left(\frac{\lambda h}{\pi w^2}\right)\right]\right\}} \tag{9}$$

They have two distinct effects on the resonance response at the waveguide input. The imaginary part $Im\{Z_S\}$ is responsible for a shift in the resonant frequency of the cavity, with respect to the empty case, while the real part $Re\{Z_S\}$ produces an amplitude reduction of the input reflection coefficient. The measurement of these two information can be, in principle, adopted to retrieve the unknown complex permittivity $\epsilon' - j\epsilon''$. However, due to the high quality factor Q of the cavity, which is equivalent to a very narrow resonant bandwidth, it is very difficult to measure the exact value of the reflection coefficient, and consequently the dielectric loss.

In alternative way, the information relative to the resonant frequencies of the empty and loaded cavity, respectively equal to f_o and f_L, can be used to retrieve the imaginary part $Im\{Z_S\}$. On the other hand, the real part $Re\{Z_S\}$ can be derived from the knowledge of the loaded dielectric quality factor Q_L, which is inversely proportional to the difference between the $3dB$ frequencies at each side of the resonance minimum. Finally, from the conjuncted knowledge of the two left terms of eqs. (8) and (9), the terms a_ϵ and b_ϵ can be retrieved, which are related to the real and imaginary parts of the unknown complex permittivity of the test surface.

The numerical implementation of the above dielectric characterization method is performed by a two step procedure, summarized as follows:

1. the imaginary part ϵ'' is first neglected and eq. (9) is solved with respect to the variable a_ϵ, directly related to the real part ϵ';

2. the value computed into step 1 is inserted into eq. (8) to retrieve the term b_ϵ, which in turns is related to the dielectric loss tangent, and thus to the imaginary part ϵ''.

3.1. K-band dielectric characterization of thin dielectric substrates

The equivalent circuit approach for the complex permittivity retrieval of planar surfaces is applied to accurately design a K-band open resonator system fed by a standard WR62 rectangular waveguide. As a primary task, the waveguide-to-cavity transition is accurately dimensioned to optimize the coupling and thus the efficiency of the open resonator system. At this purpose, the equivalent circuit approach proposed in [5] is adopted to perform a parametric analysis of the reflection coefficient at the waveguide input as a function of the waveguide height b (Fig. 3) for a fixed design frequency f_o = 24 GHz.

Due to the mechanical tolerances of the available machine, a value b_1 = 0.7 mm (Fig. 1), slightly larger than that giving the minimum reflection coefficient, is adopted for the smaller height of the transition, starting from the standard WR62 value b =4.3 mm.

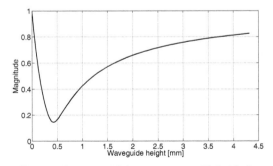

Figure 3. Reflection coefficient at the waveguide input vs. waveguide height b

The open resonator is realized on Aluminum material, by assuming a radius R_o = 517 mm and a distance l = 490 mm between the mirrors, this latter giving the excitation of a $TEM_{0,0,76}$

mode inside the cavity. The open resonator radius, the longitudinal mode number and the relative length l are derived by a compromise choice on the basis of the following parameters:

- the fixed design frequency;
- the resonator stability;
- a proper waist on the mirror to avoid diffraction losses;
- a waist value on the sample greater than a wavelength in order to measure the diffraction features of the resonant structures.

A photograph of the realized K-band open resonator system, connected to a Vector Network Analyzer, is illustrated in Fig. 4.

Figure 4. Photograph of realized K-band open resonator system

In order to test the proposed dielectric characterization technique, three standard substrates usually adopted for the realization of microwave planar structures are considered as test surfaces. The materials names and their nominal parameters as available from the producer are reported in Table 2.

Index	Material	Thickness h [mm]	ϵ' @ 10 GHz	$\tan \delta$ @ 10 GHz
1	Arlon AR600	0.762	6	0.0030
2	Arlon DiClad 527	0.762	2.55	0.0018
3	Arlon 25FR	0.762	3.58	0.0035

Table 2. Nominal parameters of test dielectric substrates

The measured return loss under empty and loaded cavity conditions are reported in Figs. 5-7. As expected from the theory outlined in the previous Section, a frequency shift and an amplitude reduction can be observed for all tested dielectrics after the introduction of the samples.

In order to retrieve the unknown complex permittivity of the substrates under test, the frequency shift $\Delta f = f_o - f_L$ is measured together with the 3 dB frequencies f_1, f_2 at each side of the resonant frequency f_L. These quantities, together with the loaded quality factor Q_L and the values of ϵ' and $\tan \delta$, as retrieved from eqs. (8)-(9) are reported in Table 3. An excellent agreement with the nominal parameters reported in Table 2 can be observed.

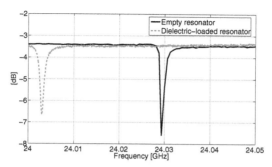

Figure 5. Measured return loss for test material 1: comparison between empty and loaded cavity conditions

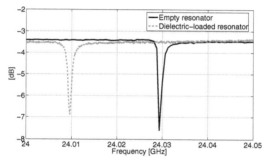

Figure 6. Measured return loss for test material 2: comparison between empty and loaded cavity conditions

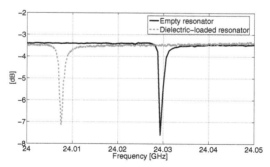

Figure 7. Measured return loss for test material 3: comparison between empty and loaded cavity conditions

Material Index	f_L [GHz]	f_1 [GHz]	f_2 [GHz]	Δf [MHz]	Q_L	ϵ'	tan δ
1	24.003	24.0018	24.0045	2.7	8890	6	0.00324
2	24.00965	24.00918	24.00995	0.77	31181	2.535	0.002
3	24.00752	24.00698	24.0084	1.42	16907	3.54	0.004

Table 3. Measured parameters and retrieved complex permittivity of test dielectric substrates

4. Open resonator application to microstrip reflectarray elements characterization

Synthesis procedures for the reflectarrays design [24] require the accurate phase characterization of the field reflected by the single radiating element, to properly choice the dimensions and distributions of the grid radiators giving a radiated field with prescribed features. The reflecting response of microstrip reflectarray elements can be characterized by inserting a periodic array of identical radiators into the open resonator system (Fig. 8). The array grid is chosen to be sufficiently large (typically greater than 5 x 5 elements) in order to better approximate the assumption of infinite array analysis [21].

The grounded reflecting surface can be modelled by the equivalent circuit of Fig. 9, where the relative parameters L_1, C_1 representing the grid depend on the variable length L of the reflectarray patches as well as on the grid spacing between adjacent elements.

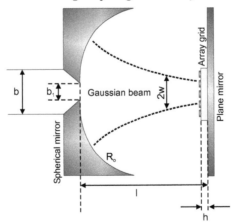

Figure 8. Open resonator system for reflectarray elements characterization

Due to the small thickness h of the usually adopted substrates (typically less than a quarter wavelength), and assuming neglecting losses in the dielectric support, an equivalent inductance L_u can be considered to model the slab, as given by the expression:

$$L_u = Z_d tan\,(k_d h - \phi_G) \qquad (10)$$

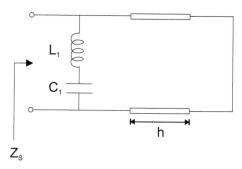

Figure 9. Equivalent circuit relevant to the grounded reflecting grid

where $Z_d = \frac{Z_0}{\sqrt{\epsilon_r}}$, $k_d = k_0\sqrt{\epsilon_r}$, ϵ_r is the relative permittivity of the dielectric and the term ϕ_G is given by eq. (6).

Under the above assumptions, the equivalent circuit of Fig. 9 simplifies as in Fig. 10.

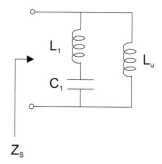

Figure 10. Simplified equivalent circuit relevant to the grounded reflecting grid

Its introduction in the original circuit of Fig. 2 affects the open resonator behavior, which consequently shows two resonant frequencies $f_{1,2}$ relevant to the circuit of Fig. 11.

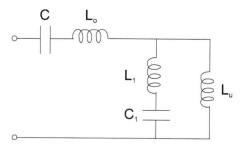

Figure 11. Resonant circuit for the computation of frequencies $f_{1,2}$

They can be easily expressed as:

$$f_{1,2} = \frac{1}{2\pi}\sqrt{\frac{c \pm \sqrt{c^2 - 4d}}{2d}} \tag{11}$$

where:

$$c = L_oC(L_1 + L_u)C_1 + CL_1L_uC_1 \tag{12}$$

$$d = L_oC + (L_1 + L_u)C_1 + L_uC \tag{13}$$

From the above equations, it is straightforward to derive a closed expression for the circuit parameters L_1, C_1, given as:

$$L_1 = \frac{C_1A_1 + B_1}{C_1E_1} \tag{14}$$

$$C_1 = \frac{B_2E_1 - B_1E_2}{A_1E_2 - A_2E_1} \tag{15}$$

where:

$$A_n = (2\pi f_n)^2 L_u - (2\pi f_n)^4 L_u L_o C \tag{16}$$

$$B_n = (2\pi f_n)^2 (L_u + L_o)C - 1 \tag{17}$$

$$E_n = (2\pi f_n)^4 (L_u + L_o)C - (2\pi f_n)^2 \tag{18}$$

for $n = 1, 2$.

The procedure for the reflecting behavior characterization of reflectarray elements can be summarized as follows:

1. insert a sufficiently large array grid of identical square patches having side L and spaced by a distance D;

2. measure the return loss at the waveguide input and derive from it the resonant frequencies $f_{1,2}$ as given by eq. (11);

3. compute the equivalent circuit parameters L_1, C_1 as given by eqs. (14), (15) to have a full characterization of the modelling circuit in Fig. 10 and finally use it to compute the reflection coefficient of the array grid.

The outlined characterization method needs to be obviously repeated for different values of the patch length L in order to retrieve the reflectarray element response as a function of the tuning geometrical parameter.

4.1. K-band phase response characterization of reflectarray elements

The K-band open resonator system of Fig. 4 is adopted to validate the reflectarray elements characterization method outlined in the previous Section. Reflectarray grids composed of 16 x 16 identical square patches with a spacing $D = 0.65\lambda_o$ are considered, for three different values of the patch length L namely $L = 3.2$ mm, 3.5 mm, 3.7 mm. A dielectric substrate oh thickness $h = 0.762$ mm and relative permittivity $\epsilon_r = 2.33$ is assumed as support. A photograph showing the open resonator system loaded with a test reflecting surface is reported in Fig. 12(a), with a particular illustrating the reflectarray grid in Fig. 12(b).

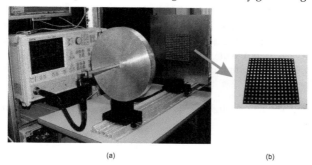

(a)   (b)

Figure 12. Photograph of open resonator system (a) loaded with the reflecting surface (b)

To derive the reflectarray element behavior, the return loss magnitude at the waveguide input is measured for the three different values of the patch side L. For all cases, as illustrated in Figs. 13-15, several resonance couples are visible, corresponding to the various modes excited into the cavity. However, the only couple to be considered is that modelled by the equivalent circuit of Fig. 2, corresponding to the $TEM_{0,0,131}$ mode, which is associated to an empty cavity resonance $f_o = 24GHz$.

In the presence of each reflecting surface, two resonant frequencies $f_{1,2}$ are produced, as highlighted in the previous section. They can be easily identified in the return loss measurements (Figs. 13-15) as follows:

i) For a patch side dimension L less than that providing the resonance condition, frequency f_1 is chosen as the nearest one (at the left side) to the resonance frequency f_o of the empty cavity, while the frequency f_2 corresponds to the resonance of the reflectarray grid, easily computed on the basis of the patch dimension L. This is the case corresponding to Fig. 13.

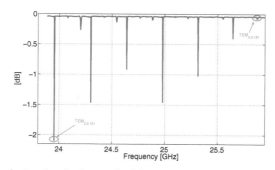

Figure 13. Measured return loss for the case $L = 3.2$ mm

ii) For a patch side dimension L equal to that providing the resonance condition, frequencies $f_{1,2}$ are chosen as those which are equally far from the resonance frequency f_o of the empty cavity. This is the case corresponding to Fig. 14.

iii) For a patch side dimension L greater than that providing the resonance condition, frequency f_1 is chosen as the nearest one (at the right side) to the resonance frequency f_o of the empty cavity, while the frequency f_2 corresponds to the resonant frequency of the reflectarray grid, again computed on the basis of the patch dimension L. This is the case corresponding to Fig. 15.

The relevant resonances are highlighted in Figs. Figs. 13-15 and summarized in Table 4.

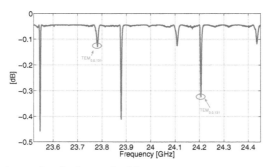

Figure 14. Measured return loss for the case $L = 3.5$ mm

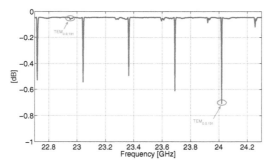

Figure 15. Measured return loss for the case $L = 3.7$ mm

Patch side L [mm]	f_1 [GHz]	f_2 [GHz]	L_1 [nH]	C_1 [fF]
3.2	23.95	25.88	4.06	7.55
3.5	23.78	24.22	4.24	8.44
3.7	22.95	24.02	3.54	8.7

Table 4. Measured resonant frequencies $f_{1,2}$ and relative circuit parameters of Fig. 10

The values of retrieved L_1, C_1 parameters are inserted into the circuit of Fig. 10 to compute the reflection phase relevant to the array grid as a function of frequency. The resulting data are successfully compared in Figs. 16-18 with results coming from Ansoft Designer simulations, for the three different dimensions of patch side L.

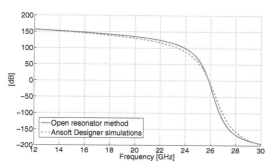

Figure 16. Reflection phase vs. frequency for the case $L = 3.2$ mm

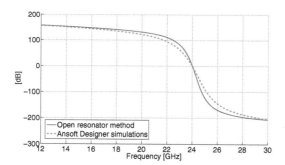

Figure 17. Reflection phase vs. frequency for the case $L = 3.5$ mm

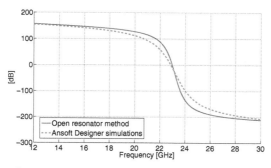

Figure 18. Reflection phase vs. frequency for the case $L = 3.7$ mm

Finally, the parametric results shown in Figs. 16-18 are combined to derive the phase design curve of the reflectarray element as a function of the tuning length L (Fig. 19). Again, a successfull comparison can be observed with Ansoft Designer simulations, thus demonstrating the effectiveness of the proposed approach.

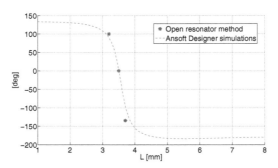

Figure 19. Reflection phase vs. patch side L

5. Conclusion

An equivalent circuit approach has been adopted in this chapter to model an open resonator system used for the equivalent impedance characterization of planar surfaces. On the basis of a full modal analysis, an accurate modelling of the open cavity is performed, taking also into account for the diffraction and the ohmic losses. This approach has led to the optimization of the coupling between the feeding waveguide and the open cavity, thus providing a significant improvement with traditional open resonator methods. The approach has been successfully applied in the framework of two specific application contexts, namely the complex permittivity retrieval of thin dielectric substrates and the phase response characterization of microstrip reflectarrays elements. In both cases, the effectiveness of the method has been experimentally demonstrated by discussing the results obtained from the application of a K-band open resonator system loaded with thin dielectric substrates and small reflectarray grids.

Author details

Sandra Costanzo, Giuseppe Di Massa and Hugo Oswaldo Moreno
University of Calabria, Italy

6. References

[1] Afsar, M. N., Birch, J. R., Clark, R. N. & Chantry, G. W. (1986). The measurement of the properties of materials. *Proc. of IEEE*, Vol. 74, (1986) page numbers (183-199).

[2] Afsar, M. N. & Button, K. J. (1985). Millimeter-wave dielectric measurement of materials. *Proc. of IEEE*, Vol. 73, (1985) page numbers (131-153).

[3] Bialkowski, M. E. & Song, H. J. (1999). Investigations into power-combining efficiency of microstrip patch transmit arrays. *Microwave Opt. Tech. Letters*, Vol. 22, (1999) page numbers (284-287).

[4] Boughriet, A.-H., Legrand, C. & Chapoton, A. (1997). Noniterative stable transmission/reflection method for low-loss material complex permittivity determination. *IEEE Trans. Microwave Theory Tech.*, Vol. 45, (1997) page numbers (52-57).

[5] Bucci, O. M. & Di Massa, G. (1992). Open resonator powered by rectangular waveguide. *IEEE Proc. H*, Vol. 139, (1992) page numbers (323-329).

[6] Costanzo, S., Venneri, I., Di Massa, G. & Borgia, A. (2010). Benzocyclobutene as substrate material for planar millimeter-wave structures: dielectric characterization and application. *Int. Journal of Infrared and Millimeter Waves*, Vol. 31, (2010) page numbers (66-77).

[7] Cullen, A. L. & Yu, P. K. (1979). Complex source-point theory of the elecromagnetic open resonator. *Proc. Royal Soc. Lond. A*, Vol. 366, (1979) page numbers (155-171).

[8] Di Massa, G., Costanzo, S. & Moreno, O. H. (2012). Accurate circuit model of open resonator system for dielectric material characterization. *J. of Electromagn. Waves and Appl.*, Vol. 26, (2012) page numbers (783-794).

[9] Djordjevic, A. R., Biljic, R. M., Likar-Smiljanic V. D. & Sarkar, T. P. (2001). Wideband frequency-domain characterization of FR-4 and time-domain causality. *IEEE Trans. Electrom. Compat.*, Vol. 43, (2001) page numbers (662-667).

[10] Dudorov, S. N., Lioubtchenko, D. V., Mallat, J. A. & Risnen, A. V. (2005). Differential open resonator method for permittivity measurements of thin dielectric film on substrate. *IEEE Trans. Microwave Theory Tech.*, Vol. 54, (2005) page numbers (1916-1920).

[11] Gui, Y. F., Dou, W. B., Su, P. G. & Yin, K. (2009). Improvement of open resonator technique for dielectric measurement at millimetre wavelengths. *IET Microwave Ant. Propag.*, Vol. 3, (2009) page numbers (1036-1043).

[12] Hasar, U. C. & Simsek, O. (2009). An accurate complex permittivity method for thin dielectric materials. *Progress In Electromagnetics Research, PIER*, Vol. 91, (2009) page numbers (123-138).

[13] Hirvonen, T. M., Vainikainen, P., Lozowski, A. & Raisanen, A. V. (1996). Measurement of dielectrics at 100 GHz wit an open resonator connected to a network analyzer. *IEEE Trans. Instrum. Measurements*, Vol. 45, (1996) page numbers (780-786).

[14] Janezic, M. D., Williams, D. F., Blaschke, V., Karamcheti, A. & Chang, C. S. (2003). Permittivity characterization of low-k thin films from transmission-line measurements. *IEEE Trans. Microwave Theory Tech.*, Vol. 51, (2003) page numbers (132-136).

[15] Jones, R. G. (1976). Precise dielectric measurements at 35 GHz using an open microwave resonator. *Proc. IEEE*, Vol. 123, (1976) page numbers (285-290).

[16] Li, K., McLean, S. J., Greegor, R. B., Parazzoli, C. G. & Tanielian, M. H. (2003). Free-space focused-beam characterization of left-handed materials. *Applied Physics Letters*, Vol. 82, (2003) page numbers (2535-2537).

[17] Linch, A. C. (1982). Measurement of permittivity by means of an open resonator. II Experimental. *Proc. Royal Soc. Lond. A*, Vol. 380, (1982) page numbers (73-76).

[18] Marsland, T. P. & Evans, S. (1987). Dielectric measurements with an open-ended coaxial probe. *IEEE Proc. H*, Vol. 134, (1987) page numbers (341-349).

[19] Napoli, L. S. & Hughes, J. J. (1971). A simple technique for the accurate determination of the microwave dielectric constant for microwave integrated circuit substrates. *IEEE Trans. Microwave Theory Tech.*, Vol. 19, (1971) page numbers (664-665).

[20] Olyphant, M. & Ball, J. H. (1970). Strip-line methods for dielectric measurements at microwave frequencies. *IEEE Trans. Electric. Insul.*, Vol. 5, (1970) page numbers (26-32).

[21] Pozar, D. M., Targonski, S. D. & Syrigos, H. D. (1997). Design of millimeter wave microstrip reflectarrays. *IEEE Trans. Antennas Propag.*, Vol. 45, (1997) page numbers (287-296).

[22] Sheen, J. (2005). Study of microwave dielectric properties measurements by various resonant techniques. *Measurement*, Vol. 37, (2005) page numbers (123-130).

[23] Thompson, D. C., Tantot, O., Jallageas, H., Ponchak, G. E., Tentzeris, M. M. & Papapolymeron, J. (2004). Characterization of liquid crystal polymer (LCP) material and transmission lines on LCP substrates from 30 to 110 GHz. *IEEE Trans. Microwave Theory Tech.*, Vol. 52, (2004) page numbers (1343-1352).

[24] Venneri, F., Costanzo, S., Di Massa, G. & Angiulli, G. (2005). An improved synthesis algorithm for reflectarrays design. *IEEE Antennas Wireless Propag. Letters*, Vol. 4, (2005) page numbers (258-261).

[25] Yu, P. K. & Cullen, A. L. (1982). Measurement of permittivity by means of an open resonator. I Theoretical. *Proc. Royal Soc. Lond. A*, Vol. 380, (1982) page numbers (49-71).

Chapter 3

Nondestructive Evaluations by Using a Prototype of a Microwave Tomograph

R. Monleone, M. Pastorino, S. Poretti, A. Randazzo,
A. Massimini and A. Salvadè

Additional information is available at the end of the chapter

1. Introduction

In the field of nondestructive testing and evaluations (NDT&E), inspection systems working at microwave frequencies are now rather common (Zoughi, 2000; Franchois et al., 1998; Heinzelmann et al., 2004; Schilz & Schiek, 1981). They usually work by considering the electromagnetic waves reflected or transmitted by a body or a specimen. The percentage of reflection or the resulting attenuation of the waves can be correlated to the characteristic of the material or to possible defects inside the structure. Recently, tomographic approaches have been proposed, too. In this case, the imaging apparatus rotates around the target under test in order to collect multi-illumination multiview data. After a proper post-processing of the measured data, slides of the target cross section are obtained and visualized as rough images of the values of the retrieved dielectric parameters. The working modality is similar to the X-ray computerized tomography, but, at this lower frequency, the scattering phenomena is significant. Consequently, two approaches can be followed. The scattering contribution can be considered as an unwanted signal to be eliminated (see, for example, (Bertero et al., 2000)) or it can be included in the retrieving process (Colton & Kress, 1998). In the last case, which is considered in the present contribution, the field measured in a given position cannot be related to the transmitted field produced by a source located at an opposite position with respect to the unknown target. On the contrary, any measured value must be related to all the sources (if more than one illuminates the target contemporarily) and, mainly, to all the scattering points inside the target itself (i.e., the entire cross section).

In the last years there has been a growing interest in developing imaging systems working at microwave frequencies both for industrial applications (Kharkovsky & Zoughi, 2007; Giakos et al., 1999; Kraszewski, 1996) and medical diagnostics (Meaney et al., 2000; Jofre et al., 1990; Hagl et al., 2003; Henriksson et al., 2010).

In this Chapter, the development of a prototype of imaging tomograph working at microwave frequencies is discussed and reviewed (Salvadè et al., 2007, Pastorino et al., 2006). By using this prototype, dielectric targets can be inspected both for material characterization as well as for nondestructive analysis. The sample under test (SUT) is positioned on a rotating platform made by weakly scattering material, which represents the investigation area. Foam microwave absorbers surround the investigation area in order to minimize reflections due to the platform itself and to the motorized antenna actuators. A transmitting antenna is used for illuminating the SUT with a known incident electric field. The scattered electric field is collected by a receiving antenna, which can rotate around the table. Both antennas can be moved vertically and are mounted on fiberglass supports. The system can collect multi-illumination multiview data at different heights by means of log periodic transmitting and receiving antennas (8.5 dBi gain). Moreover, a multi-frequency acquisition is also performed by sweeping the operating band of 1-6 GHz with steps of 166 MHz at a sweep time of 96.348 msec. The transmitted wave and the samples of the scattered electric field are obtained by using a vector network analyzer (VNA) including a digital IF input filter configured at 300 Hz. The acquisition module is controlled by a custom software, while a standard LAN connection is used to interface the VNA to the management PC. Such PC also controls and synchronizes the movements of the antennas and the platform by means of three motorized actuators, whose drivers are connected through a CAN bus.

The Chapter is organized as follows. Basic concepts about the microwave measurement methods are briefly recalled in Section 2, whereas the design and the construction of the microwave tomograph prototype are discussed in Section 3. Section 4 provides some results obtained by inverting the measured data. Finally, some conclusions are drawn in Section 5.

2. Microwave measurement methods

Microwave imaging and inspection techniques are based on the interaction between electromagnetic fields and materials as defined by Maxwell's laws (Boughriet et al.; 1997; Fratticioli, 2001; Pozar, 2005; Vincent et al. 2004; Jeffrey et al., 1996). In particular, the propagation of microwaves is influenced by material properties such as the dielectric permittivity ϵ, the electric conductivity σ, and the magnetic permeability μ. Images obtained with microwave techniques are typically maps of the distribution of these parameters within an object or a material sample. In microwave sensing applications, it is usually assumed that $\mu = \mu_0$, being μ_0 the magnetic permeability of the vacuum (nonmagnetic materials), since inspection inside metals is very difficult to achieve due to their high conductivity (thus leading to very little penetration of microwaves). In general, for lossy materials, the propagation constant is expressed as (Balanis, 1989)

$$k = \alpha + j\beta = \sqrt{j\omega\mu(\sigma + j\omega\epsilon)} \qquad (1)$$

where α and β are the attenuation and phase constants, respectively. It can be also introduced a complex permittivity, such that:

$$j\omega\epsilon_c = j\omega(\epsilon' - j\epsilon'') = \sigma + j\omega\epsilon = j\omega\epsilon\left(1 - j\frac{\sigma}{\omega\epsilon}\right) \qquad (2)$$

where the imaginary part ϵ'' accounts for the material losses. For a homogeneous material, the complex permittivity can be easily determined by measuring the electromagnetic field (amplitude and phase) at two different positions (before and after crossing the material). More in general, when the material sample is not homogeneous and contains many zones with different properties, each change of property represents a discontinuity of ϵ_r and σ, giving rise to diffractions and reflections of the electromagnetic waves.

Various measurement techniques are known and well documented in the literature (Kraszewsky, 1996; Nyfors and Vainikainen, 1989; Clarke et al., 2003) for the characterization of dielectric properties of materials. Among these, the most popular are based on transmission, reflection, and resonant sensors. Systems working with transmission methods basically consist of a transmitting and one or more receiving antennas (Fig. 1). The material sample under test is placed between the antennas and interacts with the microwave propagation by modifying its amplitude and phase. The physical parameter that describes this variation is the propagation constant k which can be obtained by measuring the changes in attenuation and phase of the microwave signal when it passes through the material.

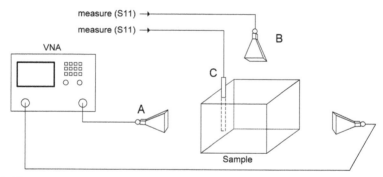

Figure 1. Microwave measurement configurations.

The measuring methods based on reflection concepts use one or more sensors that send a signal to a sample material probe and consequently analyze the amplitude and phase of the signal that is reflected back (or scattered). What is determined is the (complex) reflection coefficient Γ, which, for a normally incident plane wave impinging on an interface between two materials (e.g., the air and the unknown material) is given by (Jordan and Balmain, 1990; Balanis, 1989)

$$\Gamma = \frac{\eta_2 - \eta_1}{\eta_2 + \eta_1} = \frac{\sqrt{\epsilon_1} - \sqrt{\epsilon_2}}{\sqrt{\epsilon_1} + \sqrt{\epsilon_2}} \tag{3}$$

where η_2 and η_1 denote the intrinsic impedances of the two media (usually, medium 1 is air). When applying the reflection method, it is often useful to share the same antenna for both the transmitter and the receiver, therefore simplifying the mechanical complexity of the system. Examples of sensors based on this methodology are open-ended coaxial reflection sensors, monopole probe/antennas, open-ended waveguide sensors.

While microwave systems based on transmission and reflection concepts are now very common, systems based on microwave tomography are still rather infrequent. In a typical microwave tomograph setup, a material sample under test (SUT) is illuminated by a microwave source and the scattered field is collected by one or more receiving antennas placed at different positions around the SUT. The information of the scattered electric field is then used to obtain the dielectric properties inside the sample.

What is actually measured is the sum of the incident and the scattered fields. Since the incident field is known, the scattered field can be deduced. In order to collect as much information as possible about the scattered field, it is necessary to measure the total field at different locations. Also, the reconstruction algorithm requires that this procedure is repeated by illuminating the SUT with incident fields that arrive from different angles and directions.

The simplest configuration for a tomograph is therefore that with one moveable transmitting antenna and one moveable receiving antenna, similar to that of the transmission sensor previously described, but with the two antennas usually not facing each other. In order to reduce the amount of moving parts, an arrays of static antennas can also be used. The block diagram for the electronic hardware of a microwave tomograph with a single moveable antenna pair is shown in Fig. 2.

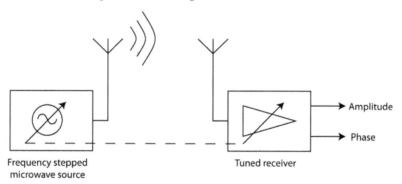

Figure 2. Simple block representation of a two-antenna tomographic set up.

The illuminating antenna is fed with a signal from a microwave generator working at one or more frequencies. The used wavelength should be chosen comparable to the electrical dimensions of the object to be analyzed. The receiver must satisfy the bandwidth requirements of the transmitter and must be able to tune to its instant frequency. In order to capture also the weakest scattered signals, the receiver should have a good sensitivity, dynamic range and selectivity. It should be possibly immune to strong interferers that may be present on other frequencies. The information sought for is the amplitude and phase of the signal picked up by the receiver antenna, compared to those of the transmitted signal. This information is typically extracted from the baseband of the receiver in the form of a complex signal.

The transmitter signal is, in many cases, a CW carrier, so that the wanted signal will be found at frequency zero (DC) at the output of the baseband block of the receiver. The receiver's task is that of converting the RF signal centered at the transmitter's frequency to baseband, with the best accuracy, dynamic range and possibly lowest noise. This can be accomplished with various receiver architectures. In the case of the superheterodyne (or heterodyne), the received RF signal, after some filtering and amplification, is first converted to an intermediate frequency (IF) by mixing (heterodyning) it with a local oscillator signal. Band (or channel) filtering is performed by a band pass filter in the receiver's IF section. Successively, the IF signal is demodulated into a base band signal by a synchronous demodulator (a mixer) or other circuits, based on the modulation type of the signal (e.g. envelope detector for an AM signal with carrier). Adequate RF filtering is therefore mandatory to block image frequencies.

For imaging applications not only the amplitude of the received signal, but also its phase is relevant. The phase information must therefore be preserved during the whole signal processing chain throughout the base band. To this end, the IF signal is converted to the base band with a vector (or I/Q) mixer as shown in Fig. 3.

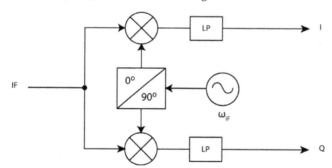

Figure 3. IF to baseband I/Q demodulator

A variant of the superheterodyne receiver aimed at mitigating the image frequency problem and used in multi-band receivers is an architecture with two consecutive IF frequency sections and mixers. The higher the chosen IF frequency, the larger the distance between the wanted RF and the unwanted RF image frequencies will be, therefore relaxing the constraint for the RF image blocking filter. On the other hand, in a multi-band receiver IF filtering is used to block adjacent channel signals. Such filtering becomes more difficult at higher IF frequencies, because the relative bandwidth of the IF filter will be smaller for a given channel width.

The double conversion receiver helps to avoid this compromise by first converting the RF signal to a higher IF (and therefore relaxing the front end filtering requirements), followed by a second converter and IF where the more demanding channel selection filtering is performed. Of the two local oscillators in a double conversion receiver, one is tunable and the other operates at a fixed frequency. A fixed frequency oscillator can be better optimized for low phase noise and is also easier to implement. This makes it appealing as the first local oscillator in some single chip receivers. In this case, the absolute bandwidth of the first IF must cover that of all the RF channels, which for some applications (wide band receivers) might not be practical. Choosing the second IF at suitably low frequency enables the opportunity to demodulate the signal to baseband with a digital I/Q demodulator (see Low IF receiver).

An alternative to actively suppress the image frequency from the mixer output in a superheterodyne receiver is offered by an image reject or Hartley mixer, where RF image signals down converted by the two mixers forming this circuit arrive at the adder with opposite phase, therefore cancelling each other. On the other hand the wanted RF signals combine in phase. The effective image frequency rejection attainable by this circuit depends strongly on the accuracy of the 90° phase shifters and is therefore limited in practical situations, since it can be quite difficult to obtain sufficiently accurate 90° shifters at higher frequency.

The architecture using low IF receivers can be considered as a variant that is halfway between the superheterodyne receiver (of which it maintains the basic block diagram) and the homodyne receiver. It maintains the advantage of the superheterodyne over the homodyne consisting of the intrinsic absence of DC offsets in the demodulated signal but it also keeps the image frequency problem which must be addressed by some of the solutions discussed earlier. The low IF frequency makes it possible to digitize the signal at IF and perform the channel filtering and baseband conversion in the digital domain, which is an interesting feature from the system integration point of view. The better accuracy of the digital I/Q demodulator also increases the receiver's SNR in presence of complex signal modulation types.

However, the various receivers suffer in general from interfering signals, both in band and out of band but to a different extent. Self-generated interference due to LO or RF signal leakage from inside or nearby the receiver can be a source of concern with direct conversion receivers, whereas interference from RF signals at the image frequency are a known potential issue with superheterodyne receivers. Distortion due to nonlinear effects inside the receiver in presence of strong interfering signals is common to all practical receivers.

The distortion related to third order intermodulation products can be an issue in any type of receiver and can be generated by some nonlinearity at any stage, including the front end amplifier. In addition, second order intermodulation products generated inside the mixer of direct conversion receivers can directly translate to baseband also producing output distortions. It is therefore important that the receiver exhibits good performance concerning this point. The above mentioned effects result in practical limitations of real receivers in terms of dynamic range. Other issues derive from the unbalance in the analog quadrature networks found in many signal blocks discussed so far and can result in amplitude and phase errors of the baseband output signal. Finally noise from the local oscillators (phase noise), baseband noise and quantization noise occurring during the analog to digital conversion can reduce the signal to noise ratio of the output signal.

A laboratory prototype of a tomograph with good performance and flexibility can be built around a network analyzer. This type of instrument contains the necessary RF signal source, the RF receiver and baseband signal processing circuitry to obtain comparative amplitude and phase measurements between the source (illuminating) signal and the received (scattered) signal. As an example to show how different types of frequency conversion and receiver architectures discussed so far can be used and combined, the following Fig. 4 shows an example of a simplified network analyzer receiver's block diagram.

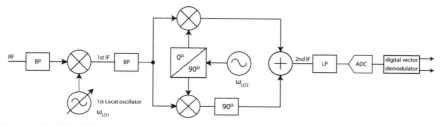

Figure 4. Simplified block diagram of the receiver section in a typical network analyser

In this dual conversion receiver, the RF signal is downconverted to a first IF frequency by mixing it with a tunable local oscillator signal. Next, after some channel filtering, the signal is converted to a second low IF by means of an image reject mixer. At this point, the signal is digitized for further processing. The vector demodulation that will produce the amplitude and phase information is performed digitally. The source (transmitting) signal generation path is usually also made of one or more frequency translation (upconversion) steps in order to cover multiple frequency bands. Transmitter and receiver must be synchronized in phase and frequency.

From the previous discussion, it is evident that more than one receiver architecture is suitable for a microwave tomograph. The choice should then be made under consideration of the environment where the instrument is to be operated (interfering signals), complexity, cost and flexibility. Direct conversion receivers are probably the most flexible, since their frequency range is basically limited only by the local oscillator tuning range and because most of the signal processing can be done digitally. On the other hand, a laboratory prototype of a tomograph with optimal performance can be realized with the help of a commercial vector network analyzer (VNA), which is often based on a dual conversion superheterodyne receiver (Agilent, 2005). It is also very important, for a microwave tomograph, that extreme care is addressed against spurious emissions and parasitic couplings between antenna cables and circuit components (the latter are particularly critical in direct reception/homodyne receivers), as they could easily exceed the amplitude of weakly scattered signals coming from the SUT, thus severely reducing the SNR of the tomograph. To ensure low emissions or couplings from the antenna cables, these should be single or double shielded and the antennas should avoid signal injection on the cable shield.

Finally, the antenna is one of the most important component for microwave imaging systems. Theoretically, many kinds of antenna can fit in the tomography setup specifications. In practical measurements, many things must be taken into account. According to the inversion algorithm needs, it is possible to select the optimal antenna characteristics that allow obtaining the best performance. In order to enhance the inversion accuracy many algorithms use multiple frequencies, some others use pulsed broadband signals. This requirement, for example, restricts the antenna choice to broadband types only. In the case of a two dimensional inversion algorithm, the SUT is assumed to be homogeneous and uniform along the z axis. In other words, the scatterer is divided into slices along this axis. Looking at the top view (Fig. 5(a)) the antenna radiation angle must be large enough to "illuminate" all the SUT surface. On the other hand, because of the properties of the inversion algorithm, the vertical radiation pattern (side view, Fig. 5(b)) must be taken as narrow as possible.

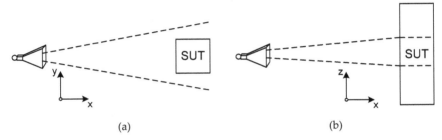

(a)   (b)

Figure 5. (a) Top and (b) side views of the SUT.

During the measurement setup design, it is also very important to consider the space surrounding the inspection area. The inversion algorithms usually assume this space to be perfectly homogeneous with known characteristics. In a real implementation this assumption is not valid. The measurement setup is composed by many mechanical components that could affect the measurements quality. A little variation of the surrounding volume can modify the measurements repeatability. The use of antennas with a limited back lobe can drastically decrease the influence of this problem. Another possibility to decrease the influence of the surrounding volume is the use of microwave absorbing materials surrounding the investigation area.

The choices of the number of antennas and their positions are extremely important. In this context there are many possible implementation options depending on the application needs and the inversion algorithm topology. One can distinguish mainly between two kinds of measurement setup: static or dynamic one. Each one brings advantages or disadvantages in relation to the specific application and the designer must therefore consider the properties of the object to be imaged. Depending on the SUT, it is generally possible to exclude right away some configurations. For example, in the case of a single piece that can fit entirely into the investigation domain, it is possible to perform a multiview inspection by rotating the SUT on its own axis instead of moving the illuminating antenna. On the other hand, if the goal is to measure only a part of a bigger object (e.g., parts of the human body, such as arms, legs or breast) the system has to perform a multiple view scanning without the SUT rotation.

3. Practical tomograph implementation example

This section describes the developed tomograph prototype (Fig. 6). In order to be as flexible as possible, a linearly polarized antenna with a large bandwidth has been chosen. Fig. 7 shows the considered antennas.

Figure 6. Prototype of the microwave tomograph (developed at SUPSI).

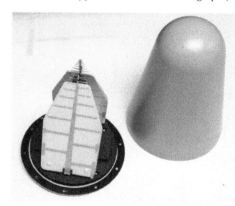

Antenna specifications	
Frequency range	850 MHz to 26.5 GHz
Polarization	Linear
Input impedance	50Ω
VSWR	≤ 2.5
Max. input power	10W to 2W
Gain	Typ. 8.5 dBi
Connector	PC 3.5 female

Figure 7. Rohde & Schwarz HL050 Antenna.

The generation, measurement and pre-processing of the microwave signal is carried out with a VNA, able to perform measurements from 10 MHz to 20 GHz. A managing software running on an external PC gathers all the measurement data through an Ethernet connection. The VNA generally requires a calibration. However, in a typical tomograph application, an empty (or void) reference measure is taken first and compared against subsequent measures with the SUT, therefore eliminating static errors due to reflections, phase adjustment errors, fixed interferers, and so on. Fig. 8 shows an example of the used VNA.

VNA Specifications	
Frequency range	10 MHz to 20 GHz
Max. output power	17dBm

Figure 8. Agilent VNA.

For mechanical reasons, some parts of the tomograph (in particular the basement) are composed by an aluminum structure. Therefore, microwave absorber material (Fig. 9 and Fig. 10) are used to prevent reflections.

Absorber specifications	
Frequency range	> 600 MHz
Reflectivity range	> -17dB
Thickness	11.4 cm

Figure 9. Eccosorb AN79

During the development of the measurement system, the planning of the illuminating signal power and receiver sensitivity play, as previously mentioned, an important role. A good dynamic range is the key for reliable and accurate measurements. The entire measurement chain can be represented with the block diagram reported in Fig. 11.

The main elements are:

- **VNA output power (typically -10 – 20 dBm):** usually the transmitted power level can be easily adjusted on the instrument. If needed, there is the possibility to add an external amplifier to increase the power level.
- **RF cable loss (typically 1 - 10 dB):** the signal amplitude loss due to the cable attenuation is variable in function of the investigation frequency band, of the cable quality and length.

- **Switching hardware loss (typically 1 - 5 dB):** the attenuation given by the switching hardware is variable and mainly depends on the board layout, the components used and the frequency band.
- **Tx/Rx antenna gain/loss (typically -2 – 10 dB of gain):** depending of the antenna type gain or loss are both possible.
- **VNA receiver sensitivity (typically -70/-100dBm):** the instrument's sensitivity level specifies the minimum possible signal amplitude required to obtain a reliable measurement. The environment noise level could also negatively affect this limit.
- **Investigation area:** the investigation area attenuation is the most critical section of the whole chain. In fact, while in every other block it is possible to correct the signal level by adding some amplification, this is not possible inside the investigation area. The inspection area is composed by the SUT and by its surrounding dielectric, which in most of the cases is air. Fig. 12 shows an example of a (rough) investigation area power budget including the antennas gain.

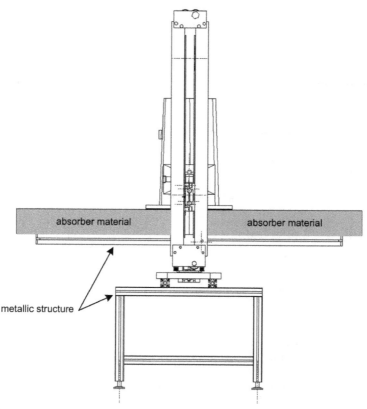

Figure 10. Absorber mounting.

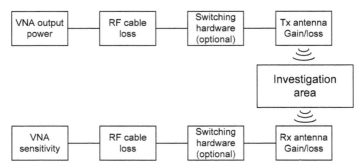

Figure 11. Measurement system block diagram

The most critical case is probably when the antennas are at a large distance and the SUT is relatively large. The specific example represents the attenuation of a SUT composed by wood ($\epsilon_r = 2.2$ and $\sigma = 0.045\ S/m$).

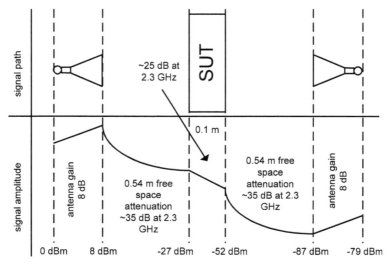

Figure 12. Simplified signal amplitude budget.

In order to acquire multiview data, in the developed tomograph prototype both the Rx antenna and the SUT are rotating and the Tx antenna is static (see the previous section). This choice leads to obtain data with the angular increments needed by the inversion algorithms. Future developments will be devoted to the implementation of a fully static design. The developed tomographic system (Fig. 6) is the result of various optimization steps. One of the major problems was the positioning accuracy of the servo motors in charge of moving the object and the Rx antenna. This accuracy should be an order of magnitude better than the required angular increments.

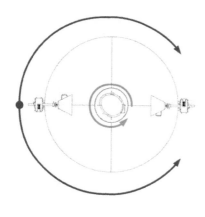

Parameters for the Rx antenna rotation:

- The Rx antenna can move from 45° to 315° with respect of the Tx antenna positioned at 0°.

- The minimum angular increment for the antenna rotation is 1°.

Parameters for the SUT rotation:

- The SUT can move 360° around its vertical axis.

- The minimum angular increment for the SUT rotation is 1°

(a)

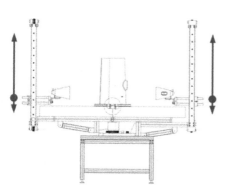

Parameters for the Rx and Tx antenna movements:

- Both antennas can move independently from 0mm to 750mm vertically.

- The minimum vertical displacement for the antennas is 1mm.

The horizontal distance of the antennas from the SUT is not under servo control but can adjusted manually. This reduces the mechanical complexity and warrants a better stability of the entire tomograph.

(b)

Figure 13. (a) Horizontal circular movement (b) and horizontal and vertical linear movement of the developed prototype.

As mentioned, in order to keep the potential reflections as low as possible, the body of the tomograph was developed in such way that all the metallic parts can easily be hidden from the inspection area using absorbers (Fig. 10). Moreover, since covering the vertical arms with absorber was unpractical, these parts have been manufactured in glass resin. These precautions have made it possible to reduce the impact of the tomograph construction on the measured samples to a strict minimum at the benefit of better results.

A major drawback of this type of configuration is the time required to obtain a single image. First of all, each motor movement (summarized in Fig. 13) requires some travel and settling time. Consequently, as the measurement points increase, the number of time intervals needed for the motors to stabilize and for the network analyzer to acquire the signals and send the data to the computer increases, too. Table 1 shows the total time required for a full series of mechanical displacements of the Rx antenna corresponding to one view of the SUT. This time needs to be multiplied by the number of different views (angular positions) of the SUT.

Number of measurement points	Angular increment (in degrees)	Time (in minutes)
3	90	0.5
11	27	1.5
16	18	2
31	9	3.5
46	6	5
91	3	10

Table 1. Time consumption based on the mechanical positioning parameters.

The acquisition software developed for the management of the tomograph and running on a PC has to communicate with two different devices:

- the servo motor control units to provide position inputs and to read back status information;
- the VNA to input acquisition settings and to transfer the measures.

All steps involved in the acquisition procedure, i.e., a precise mechanical positioning, an accurate microwave measure, a lossless data transfer and storage need to be performed sequentially because an error free execution of each of them is necessary for a successful overall result. To address these requirements a simple but robust software has been developed (Fig. 14). It consists of a linear flowchart without parallelisms, which is flexible enough to allow adding checks after every step. The simplicity of the flowchart makes it possible to easily add processing blocks or functionalities in case of future needs. For a system prototype to be used during the design and validation of new imaging algorithms this was considered an important feature.

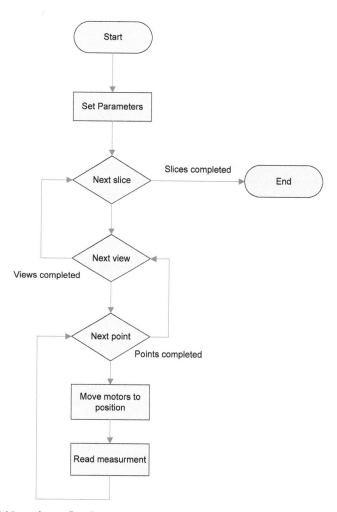

Figure 14. Acquisition software flowchart.

4. Measurement and simulation results

The measured data are stored and a processing method is applied in order to produce rough images of the reconstructed distributions of the dielectric parameters of the SUT cross section. The reconstruction procedure is based on the iterative application of the so-called distorted-Born approximation (Chew et al., 1994). Details concerning the implementation of the iterative procedures can be found in (Salvadè et al., 2010; Pastorino, 2010; Pastorino et al., 2006; Pastorino, 2004).

The capabilities of the proposed imaging systems have been preliminary assessed by means of comparisons with numerical results in terms of the measured electromagnetic field in the presence of scatters. The scatterer is a wood slab whose cross section has dimensions 11.7 × 7.8 cm. A void circular hollow of radius of 2 cm has been drilled in the sample. The values of the dielectric parameters of the wood slab are about 2.2 (relative dielectric permittivity) and 0.04 S/m (electric conductivity). Fig. 15 and Fig. 16 provides the values of the measured and simulated samples. In particular, the figure plots refer to the amplitudes and phases of the incident (Fig. 15) and total (Fig. 16) electric fields, respectively. As can be seen, the measurements are in good agreement with the simulated values.

A lot of wood and plastic slabs have been reconstructed by using the proposed prototype. The capabilities of the approach in detecting voids and defects inside these structures have been also evaluated, even in the presence of noise and interfering signals. The reader can refer to papers (Salvadè et al., 2010; Monleone et al., 2012; Pastorino et al., 2009) and the reference therein. An example is reported in the following. The inspected target is composed by a hollow wood beam with rectangular cross section of 11.5 cm × 7.5 cm and height of 50 cm (with a rectangular hole of size 5.5 cm × 3.5 cm) and a plastic object of 11 cm × 9 cm and having the same height (with a rectangular hole of size 5.3 cm × 3.0 cm) containing sand. The nominal values of the relative dielectric permittivities of wood, plastic, and sand are 1.8, 2.7, and 3, respectively. An example of the reconstruction results obtained with a frequency of 4.5 GHz is reported in Fig. 17. As can be seen, the reconstruction is fairly good and both the hollow cylinders can be quite accurately located (inside the test area) and shaped. The values of the relative dielectric permittivity are also quite similar to the actual ones, confirming that the approach is able to provide a so-called "quantitative" imaging.

More recently, the presence of metallic inclusions inside a dielectric structure has been also considered (Salvadè et al., 2008; Maffongelli et al., 2012). In this case, the imaginary part of the contrast function is retrieved, since it is related to the electric conductivity distribution of the structure. The possibility of locating metallic inclusions inside dielectric objects is clearly very appealing in several industrial application, e.g., in the wood industry, where undesired metallic object can compromise the industrial process (the row material can be not usable if foreign bodies are present). Moreover, these metallic inclusions may also damage the cutting machines used in that application.

Two examples are provided. In the first case, a wood slab of rectangular cross section of dimensions 11 cm × 9 cm is assumed. Its relative dielectric permittivity is equal to $\epsilon = 2.2$ and its electric conductivity is equal to $\sigma = 0.045\ S/m$. For the considered scenario, 3 frequencies in the range $[3, 5]\ GHz$ are assumed Moreover, $V = 8$ sources, and $M = 51$ measurement points are considered. A single inclusion, modeled as a perfect electric conducting (PEC) metallic inclusion (i.e., $\sigma = \infty$) with circular cross section of radius $r = 1\ cm$ and located at $(2.0, 0.0)\ cm$, has been considered.

The distribution of the electric conductivity $[S/m]$ provided by the inversion algorithm are shown in Fig. 18. Although, largely underestimated (no a priori information about the metallic nature of the scatterers have been included into the electromagnetic model), the distribution of the conductivity allows one to correctly identify the inclusion present in the wood slab.

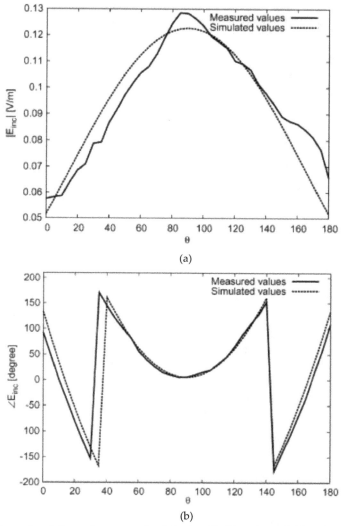

Figure 15. Comparison between the measured and simulated incident electric fields at the measurement points. (a) Amplitude (in volts per meter). (b) Phase (in degrees). © [2009] IEEE. Reprinted, with permission, from M. Pastorino, A. Salvadè, R. Monleone, G. Bozza, and A. Randazzo, "A new microwave axial tomograph for the inspection of dielectric materials," IEEE Trans. Instrum. Meas., vol. 58, no. 7, pp. 2072-2079, 2009.

In the second case, two buried objects, modeled as metallic inclusions with circular cross section of radius $r = 1\ cm$ and located at $(-2.0, 0.0)\ cm$ and $(2.0, 0.0)\ cm$, respectively, are considered. In this case, too, the reconstructed distribution of the electric conductivity, shown in Fig. 19, allows a correct identification of the two inclusions.

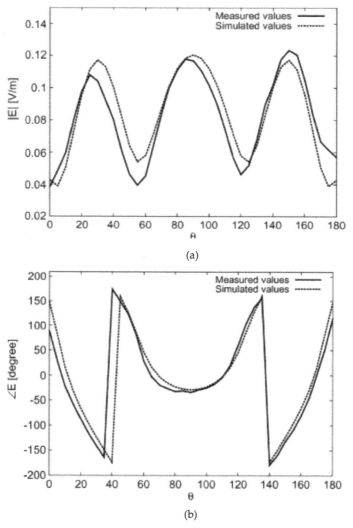

(a)

(b)

Figure 16. Comparison between the measured and simulated total electric fields at the measurement points. (a) Amplitude (in volts per meter). (b) Phase (in degrees). © [2009] IEEE. Reprinted, with permission, from M. Pastorino, A. Salvadè, R. Monleone, G. Bozza, and A. Randazzo, "A new microwave axial tomograph for the inspection of dielectric materials," IEEE Trans. Instrum. Meas., vol. 58, no. 7, pp. 2072-2079, 2009.

5. Conclusions

In this paper, the development of a prototype of a tomographic imaging systems working at microwave frequencies has been reported. Starting with a brief review of microwave measurement concepts, the details of the designed system have been reported. Morever, some measurements of the incident and total electric field have been provided, together with comparisons with simulated data. Finally, some reconstruction results concerning the

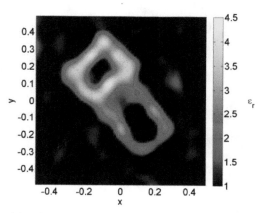

Figure 17. Reconstructed distribution of the dielectric permittivity of an inhomogeneous target (constituted by two hollow cylinders with rectangular cross sections, made of wood and plastic materials; one of them is filled by sand).

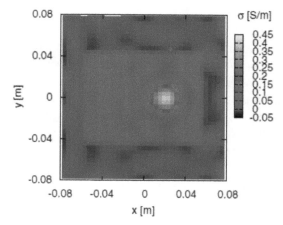

Figure 18. Reconstructed image of a wood slab with a metallic inclusion. © [2008] IEEE. Reprinted, with permission, from A. Salvadè, M. Pastorino, R. Monleone, A. Randazzo, T. Bartesaghi, G. Bozza, and S. Poretti, "Microwave imaging of foreign bodies inside wood trunks," *Proc. 2008 IEEE International Workshop on Imaging Systems and Techniques (IEEE IST08)*, Chania, Crete, Greece, Sept. 10-12, 2008.

inspection of dielectric material have been shown. The case of wood material with metallic inclusion has also been reported. Although quite preliminary, these results demonstrated that the inspection of these materials with microwave tomography is feasible. The developed system is quite simple and relatively inexpensive and can be further considered for application at several industrial levels, e.g., for nondestructive estimations in wood industry.

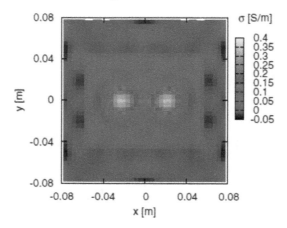

Figure 19. Reconstructed image of a wood slab with two metallic inclusions. © [2008] IEEE. Reprinted, with permission, from A. Salvadè, M. Pastorino, R. Monleone, A. Randazzo, T. Bartesaghi, G. Bozza, and S. Poretti, "Microwave imaging of foreign bodies inside wood trunks," *Proc. 2008 IEEE International Workshop on Imaging Systems and Techniques (IEEE IST08)*, Chania, Crete, Greece, Sept. 10-12, 2008.

Author details

R. Monleone, S. Poretti, A. Massimini and A. Salvadè
Department of Technology and Innovation, University of Applied Sciences of Southern Switzerland, Switzerland

M. Pastorino and A. Randazzo
Department of Naval, Electrical, Electronic and Telecommunication Engineering, University of Genoa, Italy

6. References

Agilent (2005). Appl. Note 5989-2589EN: Basic of measuring the Dielectric properties of Materials.
Boughriet, A. K.; Legrand, C. & Chapoton, A. (1997). Noniterative stable transmission/reflection method for low-loss material complex permittivity determination, IEEE Trans. Microwave Theory Tech., vol. 45 (1997), pp. 52-57.
Balanis, C. A. (1989). Advanced Engineering Electromagnetics, Wiley, NewYork, 1989.

Bertero, M.; Miyakawa, M.; Boccacci, P.; Conte, F.; Orikasa, K. & Furutani, M. (2000). Image restoration in Chirp-Pulse Microwave CT (CP-MCT), IEEE Trans. Biomed. Eng., vol. 47 (2000), pp. 690-699.

Chew, W. C.; Wang, Y. M.; Otto, G.; Lesselier, D. & Bolomey, J.-C. (1994). On the inverse source method of solving inverse scattering problems, Inv. Probl., vol. 10 (1994), 547-552.

Clarke, R. N.; Gregory, A. P.; Cannell, D.; Patrick, M-; Wylie, S.; Youngs, I.; & Hill, G. (2003). A Guide to the characterization of dielectric materials at RF and microwave frequencies, The Institute of Measurement and Control, London, 2003

Colton, D. & Kress, R. (1998). Inverse Acoustic and Electromagnetic Scattering Theory, Berlin, Springer-Verlag, 1998.

Franchois, A.; Joisel, A.; Pichot, C. & Bolomey J.C. (1998). Quantitative Microwave Imaging with a 2.45Ghz Planar Microwave Camera, IEEE Trans. Med. Imaging. vol. 17 (1998), pp. 550-561.

Fratticcioli, E. (2001). Characterization and development of a microwave industrial sensor for moisture measurements", Ph.D. Thesis, University of Perugia (Italy), 2001

Giakos, G. C.; Pastorino, M.; Russo, F.; Chowdhury, S.; Shah, N. & Davros, W. (1999). Noninvasive imaging for the new century, IEEE Instrum. Meas. Mag., vol. 2 (1999), pp. 32–35.

Hagl, D. M.; Popovic, D.; Hagness, S. C., Booske, J. H. & Okoniewski, M. (2003). Sensing volume of open-ended coaxial probes for dielectric characterization of breast tissue at microwave frequencies, IEEE Trans. Microwave Theory Tech., vol. 51 (2003), pp. 1194–1206.

Heinzelmann, E.; Salvadè, A. & Monleone, R. (2004). Wavetester, die nicht invasive Schnüffelnase, Tech. Rundsch., vol. 17 (2004), p. 50.

Henriksson, T.; Joachimowicz, N.; Conessa, C. & Bolomey, J.-C. (2010). Quantitative microwave imaging for breast cancer detection using a planar 2.45 GHz system, IEEE Trans. Instrum. Meas., vol. 59 (2010), pp. 2691-2699.

Kharkovsky, S. & Zoughi, R. (2007). Microwave and millimeter wave nondestructive testing and evaluation—Overview and recent advances, IEEE Instrum. Meas. Mag., vol. 10 (2007), pp. 26–38.

Kraszewski (1996). Microwave Aquametry, IEEE press, 1996.

Jeffrey, A.; Jargon & Janezic, M. D. (1996). Measuring complex permittivity and permeability using time domain network analysis, Proc. IEEE Int. Microwave Symp., 17-21, June 1996, vol. 3, pp. 1407-1410.

Jofre, L.; Hawley, M.; Broquetas, A.; de los Reyes, E.; Ferrando, M. & Elias-Fuste, A. (1990). Medical imaging with a microwave tomographic scanner, IEEE Trans. Biomed. Eng., vol. 37 (1990), pp. 303 –312.

Jordan, E. C. & Balmain, K. G. (1990). Electromagnetic waves and radiating systems, PrenticeHall, Second Ed.

Maffongelli, M.; Monleone, R.; Randazzo, A.; Pastorino, M.; Poretti, S. & Salvadè, A. (2012). Experimental reconstructions of dielectric targets with metallic inclusions by microwave imaging, Proc. Advanced Electromagnetics Symposium 2012 (AES'2012), Paris, France, April 16-19, 2012.

Meaney, P. M.; Fanning, M. W.; Li, D.; Poplack, S. P. & and Paulsen, K. D. (2000). A clinical prototype for active microwave imaging of the breast, IEEE Trans. Microwave Theory Tech., vol. 48 (2000), pp. 1841–1853.

Monleone, R.; Pastorino, M.; Fortuny-Guasch, J.; Salvadè, A.; Bartesaghi, T.; Bozza, G.; Maffongelli, M.; Massimini, A. & Randazzo, A. (2012). Impact of background noise on dielectric reconstructions obtained by a prototype of microwave axial tomograph, IEEE Trans. Instrum. Meas., vol. 61, no. 1 (2012), pp. 140-148.

Nyfors, E. & Vainikainen, P. (1989). Industrial Microwave Sensors, Artech House, 1989.

Pastorino, M. (2004). Recent inversion procedures for microwave imaging in biomedical, subsurface detection and nondestructive evaluation, Measurement, vol. 36, no. 3/4 (Oct.–Dec. 2004.), pp. 257–269.

Pastorino, M. (2010) Microwave Imaging, Wiley, Hoboken, 2010.

Pastorino, M.; Salvadè, A.; Monleone, R.; Bozza, G. & Randazzo, A. (2009). A new microwave axial tomograph for the inspection of dielectric materials, IEEE Trans. Instrum. Meas., vol. 58, no. 7 (2009), pp. 2072-2079.

Pastorino, M.; Salvadè, A.; Monleone, R. & Randazzo, A. (2006). A Microwave Axial Tomograph: Experimental Set Up and Reconstruction Procedure, Proc. IEEE 2006 Instrum. Meas. Technol. Conf., Sorrento, Italy, Apr. 24–27, 2006, pp. 392–396.

Pozar, D. (2005). Microwave Engineering, Third Ed., Wiley, 2005.

Salvadè, A.; Pastorino, M.; Monleone, R.; Randazzo, A.; Bartesaghi, T. & Bozza, G. (2007). A Non-Invasive Microwave Method for the Inspection of Wood Beams, Proc. 3rd Int. Conf. Electromagn. Near-Field Characterization Imag. (ICONIC), St. Louis, MO, Jun. 27–29, 2007, pp. 395–400.

Salvadè, A.; Pastorino, M.; Monleone, R.; Randazzo, A.; Bartesaghi, T.; Bozza, G. & Poretti, S. (2008). Microwave imaging of foreign bodies inside wood trunks, Proc. 2008 IEEE International Workshop on Imaging Systems and Techniques (IEEE IST08), Chania, Crete, Greece, Sept. 10-12, 2008.

Salvadè, A.; Pastorino, M.; Monleone, R.; Bozza, G.; Bartesaghi, T.; Maffongelli, M. & Massimini, A. (2010). Experimental evaluation of a prototype of a microwave imaging system, Proc. 2010 URSI Commission B International Symposium on Electromagnetic Theory, Berlin, Germany, August 16-19, 2010, pp. 1108-1111.

Schilz, W. & Schiek, B. (1981). Microwave Systems for Industrial Measurements, Adv. Electron. Electron Phys., vol. 55 (1981), pp. 309–381.

Vincent, D.; Jorat, L., & Noyel G. (2004). Improvement of the transmission/reflection method for dielectric and magnetic measurements on liquids between 0.1 and 20 GHz, Meas. Sci. Technol., vol. 5 (2004), pp. 990-995.

Zoughi, R. (2000). Microwave Nondestructive Testing and Evaluation, Kluwer, Dordrecht, The Netherlands, 2000.

Chapter 4

Non-Invasive Microwave Characterization of Dielectric Scatterers

Sandra Costanzo, Giuseppe Di Massa,
Matteo Pastorino, Andrea Randazzo and Antonio Borgia

Additional information is available at the end of the chapter

1. Introduction

Microwave tomography is a technique aimed at inspecting unknown bodies by using an incident radiation generated at microwave frequencies (Ali & Moghaddam, 2010; Bellizzi, Bucci, & Catapano, 2011; Catapano, Crocco, & Isernia, 2007; Chen, 2008; Ferraye, Dauvignac, & Pichot, 2003; Gilmore, Mojabi, & LoVetri, 2009; Habashy & Abubakar, 2004; Isernia, Pascazio, & Pierri, 2001; Kharkovsky & Zoughi, 2007; Lesselier & Bowler, 2002; Litman, Lesselier, & Santosa, 1998; Oliveri, Lizzi, Pastorino, & Massa, 2012; Pastorino, 2010; Rekanos, 2008; Schilz & Schiek, 1981; Shea, Kosmas, Van Veen, & Hagness, 2010; Zhang & Liu, 2004; Zhou, Takenaka, Johnson, & Tanaka, 2009). An illuminating system (Costanzo & Di Massa, 2011; Paulsen, Poplack, Li, Fanning, & Meaney, 2000; Zoughi, 2000) is used to produce the incident waves that interact with the body to produce a scattered electromagnetic field. Another system is used to acquire the measurements that are used as input values for the reconstruction procedures. These values are the field samples resulting from the sum of the incident and scattered waves. Since the incident field (i.e., the field produced by the illuminating system when the object is not present) is a known quantity, the scattered electric field can be obtained by a direct subtraction. Moreover, the scattering field is related to the properties of the unknown body by well-known key relationships. In particular, both position, shape and dielectric parameters of the target affect the scattered field. In this Chapter, we consider the inspection of (possibly inhomogeneous) dielectric targets, which are characterized by the distributions of the dielectric permittivity and electric conductivity, whereas magnetic materials (e.g., materials for which the magnetic permittivity is different from the vacuum one) are not considered (El-Shenawee, Dorn, & Moscoso, 2009; Franchois & Pichot, 1997). The relationship between the target properties and the sampled scattered electric field is, in integral form, a Fredholm equation of the first kind, usually indicated as the "data equation" (Bucci, Cardace, Crocco, & Isernia, 2001; Rocca, Benedetti, Donelli,

Franceschini, & Massa, 2009; van den Berg & Abubakar, 2001). The kernel of this equation is the Green's function for two dimensional geometries in free space (in this Chapter we consider imaging configurations in free space, although other configurations - e.g., half space imaging - could be assumed by introducing the proper Green's function for the specific configuration). The considered problem belongs to the class of inverse problems, which are usually ill-posed, in the sense that the solution can be not unique and unstable. To face the ill-posedness of the problem, the "data equation" is often solved together with the so-called "state equation", relating the incident electric field inside the inspected object to the problem unknowns. In particular, in the developed approach, the two equations are combined together and a single nonlinear equation is obtained. In order to numerically solve the inverse problem, a discretization is usually necessary. A pixelated image of the scattering cross section can be obtained by using square pulse basis functions. The discrete nature of the measurements (we assume that each measurement antenna is able to collect the field at a given point inside a fixed observation domain) is equivalent to consider Dirac delta functions as testing function. The result of the discretization is a (nonlinear) system of equations to be solved, usually very ill-conditioned. In order to solve, in a regularized sense, the inverse scattering problem in the discrete setting, an iterative algorithm based on an inexact-Newton method is applied (Bozza, Estatico, Pastorino, & Randazzo, 2006; Estatico, Bozza, Massa, Pastorino, & Randazzo, 2005).

The reconstruction method proposed in this Chapter can be, in principle, applied to a large variety of dielectric objects, having homogeneous or multilayer cross-sections with arbitrary shape. Only for demonstration purpose, a simple homogeneous reference target of known dielectric properties is assumed in the following. The scattered field is acquired, both in amplitude and phase, on a square investigation domain around the target, sufficiently extent to be within the radiating near-field region (Costanzo & Di Massa, 2011). The incident field, oriented along the cylindrical target axis, is produced by a standard horn antenna, and a probe of the same kind is used to collect the field on the acquisition domain, for different positions of the illuminating horn. The measured scattered field data are subsequently processed to solve the inverse scattering problem and successfully retrieve the dielectric profile of the target under test.

The Chapter is organized as follows. In Section 2, a detailed mathematical description of the reconstruction method and the relative solving procedure are provided. The imaging setup configuration and the performed scattering measurements are discussed in Section 3. Some preliminary results concerning the inversion of measured data are reported in Section 4. Finally, conclusions are outlined in Section 5.

2. Mathematical formulation

The considered approach assumes tomographic imaging conditions and it aims at reconstructing the distributions of the dielectric properties of a slice of the target (Fig. 1). A transmitting (TX) antenna is successively positioned in S different locations $\mathbf{r}_s^{TX}$, $s = 1, ..., S$, and generates a set of known z-polarized incident waves, whose electric field vectors can be expressed as:

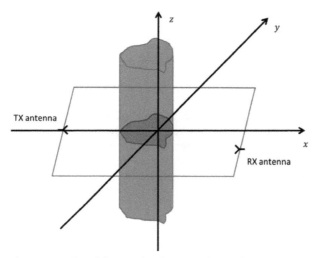

Figure 1. Schematic representation of the considered tomographic configuration

$$\mathbf{E}^s_{inc}(\mathbf{r},\omega) = e^s_{inc}(x,y,\omega)\hat{\mathbf{z}} \qquad (1)$$

where ω is the angular working frequency. It should be noted that the proposed approach does not require plane wave illumination of the target.

The object is assumed to have cylindrical geometry, with the cylindrical axis directed along a direction parallel to the electric field vectors of the incident electric field (i.e., the z axis). Moreover, the dielectric properties are assumed to be independent from the z coordinate, i.e., $\epsilon(\mathbf{r}) = \epsilon(x,y)$ and $\sigma(\mathbf{r}) = \sigma(x,y)$, being ϵ and σ the dielectric permittivity and the electric conductivity, respectively.

The object interacts with the impinging electric field. From the above hypotheses it results that the resulting total electric field is z-polarized, too, and it can be written as:

$$\mathbf{E}^s_{tot}(\mathbf{r},\omega) = e^s_{tot}(x,y,\omega)\hat{\mathbf{z}} = e^s_{scatt}(x,y,\omega)\hat{\mathbf{z}} + e^s_{inc}(x,y,\omega)\hat{\mathbf{z}} \qquad (2)$$

where $e^s_{scatt}(x,y,\omega)$ is the *scattered* electric field (due to the sth illumination), which is a mathematical quantity taking into account for the interaction effect between the incident electric field and the target. The total electric field is measured, for any location of the transmitting antenna, by a receiving (RX) antenna successively positioned in M points $\mathbf{r}^{RX}_{s,m}$, $s = 1, ..., S$, $m = 1, ..., M$.

From a mathematical point of view, the relationship between the measured total electric field and the dielectric properties of the target can be modeled by using a Lippmann-Schwinger equation (Pastorino, 2010), i.e.,

$$e^s_{tot}(x,y,\omega) = e^s_{inc}(x,y,\omega) + j\frac{k_0^2}{4}\int_D \tau(x',y')\, e^s_{tot}(x',y',\omega)H_0^{(2)}(k_0\rho)dx'dy', \; s = 1, ..., S \quad (3)$$

where $k_0 = \omega\sqrt{\epsilon_0\mu_0}$ is the free-space wavenumber (being ϵ_0 and μ_0 the dielectric permittivity and the magnetic permeability of the vacuum, respectively). In equation (3), $H_0^{(2)}$ is the Hankel function of zero-th order and second kind, $\rho = \sqrt{(x-x')^2 + (y-y')^2}$, and τ denotes the *contrast function*, which is defined as:

$$\tau(x,y) = \epsilon(x,y) - \frac{j\sigma(x,y)}{\omega\epsilon_0} - 1 \tag{4}$$

The imaging procedure, starting from the measured samples of the total electric field in the $S \times M$ measurement locations, is aimed at retrieving the contrast function, which contains all information about the unknown distributions of the dielectric properties. Consequently, equation (3) is discretized (by using pulse basis function to represent the unknowns (Richmond, 1965)) and computed at the measurement positions, leading to the following set of discrete equations:

$$\mathbf{e}_{scatt}^s = \mathbf{H}^s \mathrm{diag}(\boldsymbol{\tau})\mathbf{e}_{tot}^s, \, s = 1, \dots, S \tag{5}$$

where:

$$\mathbf{e}_{scatt}^s = \begin{bmatrix} e_{scatt}^s\left(x_{s,1}^{RX}, y_{s,1}^{RX}, \omega\right) \\ \vdots \\ e_{scatt}^s\left(x_{s,M}^{RX}, y_{s,M}^{RX}, \omega\right) \end{bmatrix}, \, s = 1, \dots, S \tag{6}$$

is an array containing the samples of the scattered electric field at the M measurement points (being $\left(x_{s,m}^{RX}, y_{s,m}^{RX}\right)$, $m = 1, \dots, M$, their positions in the transverse plane) for the sth illumination, the vector:

$$\mathbf{e}_{tot}^s = \begin{bmatrix} e_{tot}^s(x_1^D, y_1^D, \omega) \\ \vdots \\ e_{tot}^s(x_N^D, y_N^D, \omega) \end{bmatrix} \tag{7}$$

is an array containing the values of total electric field in the centers (x_i^D, y_i^D), $i = 1, \dots, N$, of the N subdomains used to discretize the investigation area, and:

$$\boldsymbol{\tau} = \begin{bmatrix} \tau(x_1^D, y_1^D) \\ \vdots \\ \tau(x_N^D, y_N^D) \end{bmatrix} \tag{8}$$

is an array containing the coefficients of the discretized contrast function. Finally, the matrix $\mathbf{H}^s$ is given by:

$$\mathbf{H}^s = j\frac{k_0^2}{4}\begin{bmatrix} h_{s,1,1} & \cdots & h_{s,1,N} \\ \vdots & \ddots & \vdots \\ h_{s,M,1} & \cdots & h_{s,M,N} \end{bmatrix}, \, s = 1, \dots, S \tag{9}$$

whose elements are provided by the following relation:

$$h_{s,m,i} = \int_{D_i} H_0^{(2)}\left(k_0\rho_{s,m}\right)dx'dy' \tag{10}$$

being D_i the ith subdomain and $\rho_{s,m} = \sqrt{\left(x_{s,m}^{RX} - x'\right)^2 + \left(y_{s,m}^{RX} - y'\right)^2}$.

It is worth noting that the array $\mathbf{e}_{tot}^s$ is unknown into equation (5). Consequently, a second equation is needed to solve the inverse problem. Such relation is found by applying (3) to points inside the investigation area. By using the same discretization, we obtain the following matrix relation:

$$\mathbf{e}_{tot}^s = \mathbf{e}_{inc}^s - \mathbf{G}\text{diag}(\tau)\mathbf{e}_{tot}^s, \quad s = 1, \ldots, S \tag{11}$$

where:

$$\mathbf{e}_{inc}^s = \begin{bmatrix} e_{inc}^s(x_1^D, y_1^D, \omega) \\ \vdots \\ e_{inc}^s(x_N^D, y_N^D, \omega) \end{bmatrix}, \quad s = 1, \ldots, S \tag{12}$$

is an array containing the values of the incident electric field in the centers of the N subdomains D_i. Moreover, the term:

$$\mathbf{G} = j\frac{k_0^2}{4} \begin{bmatrix} g_{1,1} & \cdots & g_{1,N} \\ \vdots & \ddots & \vdots \\ g_{N,1} & \cdots & g_{N,N} \end{bmatrix} \tag{13}$$

is a matrix whose elements are given by:

$$g_{i,k} = \int_{D_k} H_0^{(2)}(k_0\rho_i)dx'dy' \tag{14}$$

with $\rho_i = \sqrt{\left(x_i^D - x'\right)^2 + \left(y_i^D - y'\right)^2}$.

Equations (5) and (11) are combined together in order to obtain the following set of nonlinear equations:

$$\mathbf{e}_{scatt}^s = \mathbf{H}\text{diag}(\tau)\left(\mathbf{I} - \mathbf{G}^s\text{diag}(\tau)\right)^{-1}\mathbf{e}_{inc}^s = \mathbf{A}^s(\tau), \quad s = 1, \ldots, S \tag{15}$$

which can be written as:

$$\mathbf{e}_{scatt} = \begin{bmatrix} \mathbf{e}_{scatt}^1 \\ \vdots \\ \mathbf{e}_{scatt}^S \end{bmatrix} = \begin{bmatrix} \mathbf{A}^1(\tau) \\ \vdots \\ \mathbf{A}^S(\tau) \end{bmatrix} = \mathbf{A}(\tau) \tag{16}$$

Equation (16) needs to be solved in order to retrieve the contrast function τ. Once this term is obtained, the distributions of the dielectric parameters can be calculated using equation (4). The solution of equation (16) represents however a highly ill-posed problem. Consequently, a regularized inversion algorithm must be used (Autieri, Ferraiuolo, & Pascazio, 2011; Lobel, Blanc-Féraud, Pichot, & Barlaud, 1997). The solving procedure is based on a two-step iterative strategy (Bozza et al., 2006; Estatico et al., 2005), in which an outer linearization is performed by means of an Inexact-Newton scheme and a regularized solution to the obtained linear system is calculated by means of a truncated Landweber algorithm (Landweber, 1951).

The developed iterative procedure works as follows.

1.   Set iteration index to $n = 0$ and initialize the unknown τ_n, e.g., by choosing $\tau_n = 0$;
2.   Linearize the equation $\mathbf{e}_{scatt} = \mathbf{A}(\tau)$ in order to obtain a linear equation $\mathbf{J}_n\mathbf{h} = \mathbf{e}_n$, where $\mathbf{e}_{scatt} - \mathbf{A}(\tau_n)$ and $\mathbf{J}_n$ is the Jacobian matrix (i.e., the discrete counterpart of the Frechét derivative) of $\mathbf{A}$ at point τ_n, which is given by (Remis & van den Berg, 2000):

$$\mathbf{J}_n = \begin{bmatrix} \mathbf{J}_n^1 \\ \vdots \\ \mathbf{J}_n^S \end{bmatrix} = \begin{bmatrix} \mathbf{H}_n^1 \mathrm{diag}\big(\mathbf{e}_{tot_n}^1\big) \\ \vdots \\ \mathbf{H}_n^S \mathrm{diag}\big(\mathbf{e}_{tot_n}^S\big) \end{bmatrix} \tag{17}$$

being $\mathbf{e}_{tot_n}^s = \big(\mathbf{I} - \mathbf{G}\mathrm{diag}(\tau_n)\big)^{-1}\mathbf{e}_{inc}^s$ the total electric field (for the sth illumination) inside the investigation area due to the current estimate of the solution τ_n and $\mathbf{H}_n^s$ a inhomogeneous Green matrix given by $\mathbf{H}_n^s = \big(\mathbf{I} - \mathbf{G}\mathrm{diag}(\tau_n)\big)^{-1}\mathbf{H}^s$;

3.   Find a regularized solution $\mathbf{h}$ of the linearized equation by using a truncated Landweber algorithm;
4.   Update current solution with $\tau_{n+1} = \tau_n + \mathbf{h}$;
5.   Check if a convergence criteria (e.g., a maximum number of iterations n_{max} or a threshold on the residual) is fulfilled. Otherwise go back to step 2.

The truncated Landweber algorithm can be summarized as follows.

1.   Set iteration index to $l = 0$ and initialize the unknown $\mathbf{h}_l = 0$;
2.   Update the current solution with

$$\mathbf{h}_{l+1} = \mathbf{h}_l - \beta\mathbf{J}_n^*(\mathbf{J}_n\mathbf{h}_l - \mathbf{e}_n) \tag{18}$$

where $\beta = 0.5/\|\mathbf{J}_n^*\mathbf{J}_n\|^2$ and $\mathbf{J}_n^*$ is the adjoint of $\mathbf{J}_n$;
3.   Check if a convergence criteria (e.g., a maximum number of iterations, l_{max}, or a threshold on the residual) is fulfilled. Otherwise go back to step 2.

3. Imaging setup configuration and measurements

In order to prove the effectiveness of the approach, experimental validations are performed on a reference target by adopting the setup configuration in Fig. 2. A square investigation domain D around the scatterer is assumed, with a transmitting antenna giving an incident field oriented along the longitudinal axis of the target and assuming the positions indicated in Fig. 2 as black circles. A receiving antenna is adopted to collect the complex (amplitude and phase) field scattered by the test object on the void circles, at spacings Δx, Δy satisfying the Shannon's sampling theorem (Bucci & Franceschetti, 1989).

3.1. X-band measurements

The imaging configuration described in Fig. 2 is realized into the anechoic chamber of Microwave Laboratory at University of Calabria. Two standard X-band horn antennas are adopted as transmitting and receiving antennas, as providing a sufficiently large pattern impinging on the reference target. This is given by a wood cylinder of length equal about to 60 cm (~ 20 λ @ 10 GHz) and square cross section of side equal about to 4 cm. A photograph showing the imaging setup is reported in Fig. 3.

Scattered field measurements are performed at a frequency equal to 10 GHz, on a square measurement domain of side equal to $39\ cm$, which has been discretized into $4M$ points ($M = 53$) with spacings $\Delta x = \Delta y = \lambda/4$. As highlighted above, both transmitting and receiving horns are oriented with the field parallel to the cylinder axis. The amplitude and phase behavior of the measured scattered field is reported in Figs. 4-7 for two different positions of the illuminating horn, along two different sides of the acquisition domain scanned by the receiving probe. The positions of transmitting and receiving antennas are visible in the picture within the same figures. Similar behaviors are obtained for the other positions of the transmitting and the receiving antennas.

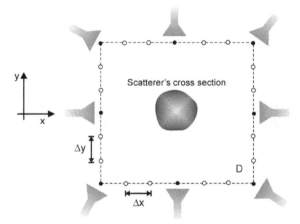

Figure 2. Imaging setup configuration

Figure 3. Photograph of measurement setup into the Microwave Laboratory at University of Calabria

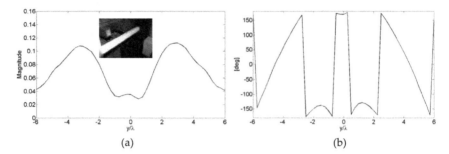

Figure 4. Amplitude (a) and phase (b) of measured scattered field: configuration in the picture

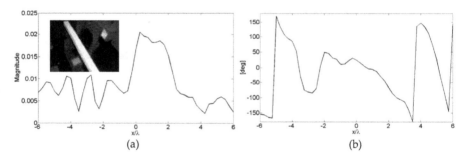

Figure 5. Amplitude (a) and phase (b) of measured scattered field: configuration in the picture

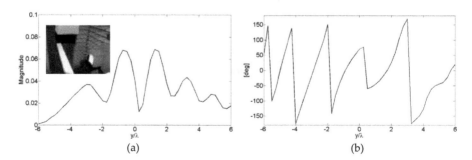

Figure 6. Amplitude (a) and phase (b) of measured scattered field: configuration in the picture

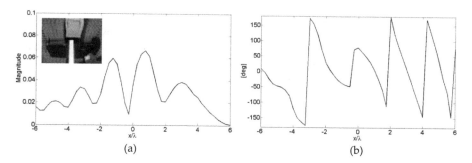

Figure 7. Amplitude (a) and phase (b) of measured scattered field: configuration in the picture

4. Preliminary reconstruction results

In this section, preliminary reconstruction results are reported. Figure 8 provides the reconstructed image of the object described in Section 3. It is obtained by inverting the real measured data described in Section 3 by using the procedure described in Section 2.

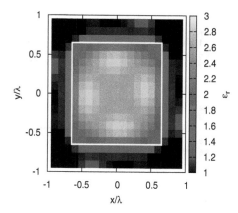

Figure 8. Reconstructed distribution of the relative dielectric permittivity inside the square investigation domain

For every side of the measurement domain, three positions of the TX antenna are used. In particular, for the first side (i.e., the one characterized by coordinate $x = -19.5\ cm$), the y-positions of TX antenna are equal to -3.625, 0, and $3.625\ cm$. For these three source positions, only the M measurement points located on the opposite side of the measurement domain (i.e., for the first side, those characterized by coordinate $x = 19.5\ cm$) are used. The remaining views are constructed in a similar way. The total number of illuminations is $S = 12$ and the total number of measured samples is $S \times M = 636$. The investigation area is assumed of square shape and side equal to 0.06 m. It is partitioned into 20×20 square subdomains. The algorithm is initialized by using a rough estimate obtained by means of a back-propagation algorithm (Lobel, Kleinman, Pichot, Blanc-Feraud, & Barlaud, 1996). The inner loop (Landweber algorithm) is stopped after a fixed number of iterations $l_{max} = 5$. The outer loop (Newton linearization) is stopped according to the L-curve criteria (Vogel, 2002), leading to an estimated optimal number of outer iterations of $n_{lc} = 3$. Maximum, minimum and mean values of the retrieved dielectric permittivity distribution are reported in Table 1. Finally, Fig. 9 reports the profiles obtained by cutting the 2D distribution along two horizontal and vertical axes passing from the center of the investigation domain. As can be seen, the presence of the target can be suitably retrieved.

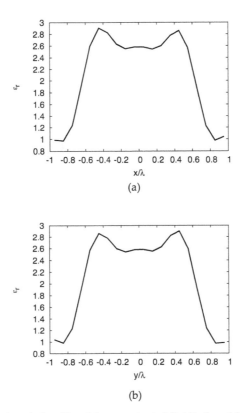

Figure 9. Horizontal and vertical profiles of the reconstructed distribution of the relative dielectric permittivity along lines passing from the center of the investigation domain

	Min	Max	Mean	Variance
Whole domain	1.00	2.90	1.67	0.48
Object	1.10	2.90	2.30	0.20
Background	1.00	1.41	1.08	0.01

Table 1. Values of the retrieved relative dielectric permittivity

5. Conclusion

The non-invasive inspection of dielectric objects has been considered in this Chapter to provide an accurate characterization of the permittivity profile at microwave frequencies. A mathematical formulation in terms of a Fredholm integral equation of the first order has been assumed, and a suitable discretization has been performed in order to numerically solve the resulting inverse problem, with a regularization approach adopted to overcome the intrinsic ill-posedness. The proposed imaging technique has been experimentally assessed by performing scattered field measurements on a square investigation domain surrounding a cylindrical dielectric target of known properties. Measured X-band data acquired by a standard horn antenna have been collected for different positions of the illuminating horn, and a successful reconstruction of the expected dielectric profile has been obtained from the application of the proposed technique.

Author details

Sandra Costanzo, Giuseppe Di Massa and Antonio Borgia
University of Calabria, Italy

Matteo Pastorino and Andrea Randazzo
University of Genoa, Italy

6. References

Ali, M. A., & Moghaddam, M. (2010). 3D Nonlinear Super-Resolution Microwave Inversion Technique Using Time-Domain Data. *IEEE Trans. Antennas Propag.*, Vol. 58, (2010) page numbers (2327–2336).

Autieri, R., Ferraiuolo, G., & Pascazio, V. (2011). Bayesian Regularization in Nonlinear Imaging: Reconstructions From Experimental Data in Nonlinearized Microwave Tomography. *IEEE Trans. Geosci. Remote Sens.*, Vol. 49, (2011) page numbers (801–813).

Bellizzi, G., Bucci, O. M., & Catapano, I. (2011). Microwave Cancer Imaging Exploiting Magnetic Nanoparticles as Contrast Agent. *IEEE Trans. Biomed. Eng.*, Vol. 58, (2011), page numbers (2528–2536).

Bozza, G., Estatico, C., Pastorino, M., & Randazzo, A. (2006). An Inexact Newton Method for Microwave Reconstruction of Strong Scatterers. *IEEE Antennas Wirel. Propag. Lett.*, Vol. 5, (2006) page numbers (61–64).

Bucci, Cardace, N., Crocco, L., & Isernia, T. (2001). Degree of Nonlinearity and A New Solution Procedure in Scalar Two-Dimensional Inverse Scattering Problems. *J. Opt. Soc. Am. A*, Vol. 18, (2001) page numbers (1832–1843).

Bucci, & Franceschetti, G. (1989). On the Degrees of Freedom of Scattered Fields. *IEEE Trans. Antennas Propag.*, Vol. 37, (1989) page numbers (918–926).

Catapano, I., Crocco, L., & Isernia, T. (2007). On Simple Methods for Shape Reconstruction of Unknown Scatterers. *IEEE Trans. Antennas Propag.*, Vol. 55, (2007) page numbers (1431-1436).

Chen, X. (2008). Signal-Subspace Method Approach to the Intensity-Only Electromagnetic Inverse Scattering Problem. *J. Opt. Soc. Am. A*, Vol. 25, (2008) page numbers (2018–2024).

Costanzo, S. & Di Massa, G. (2011). Advanced numerical techniques for near-field antenna measurements, In: *Numerical simulations of physical and engineering processes*, J. Awrejcewicz Ed., pp. (321-338), INTECH, ISBN 978-953-307-620-1, Croatia

El-Shenawee, M., Dorn, O., & Moscoso, M. (2009). An Adjoint-Field Technique for Shape Reconstruction of 3-D Penetrable Object Immersed in Lossy Medium. *IEEE Trans. Antennas Propag.*, Vol. 57, (2009) page numbers (520–534).

Estatico, C., Bozza, G., Massa, A., Pastorino, M., & Randazzo, A. (2005). A Two-Step Iterative Inexact-Newton Method for Electromagnetic Imaging of Dielectric Structures from Real Data. *Inv. Probl.*, Vol. 21, (2005) page numbers (S81–S94).

Ferraye, R., Dauvignac, J.-Y., & Pichot, C. (2003). An Inverse Scattering Method Based on Contour Deformations by Means of a Level Set Method Using Frequency Hopping Technique. *IEEE Trans. Antennas Propag.*, Vol. 51, (2003) page numbers (1100–1113).

Franchois, A., & Pichot, C. (1997). Microwave imaging-complex permittivity reconstruction with a Levenberg-Marquardt method. *IEEE Trans. Antennas Propag.*, Vol. 45, (1997), page numbers (203–215).

Gilmore, C., Mojabi, P., & LoVetri, J. (2009). Comparison of an Enhanced Distorted Born Iterative Method and the Multiplicative-Regularization Contrast Source Inversion method. *IEEE Trans. Antennas Propag.*, Vol. 57, (2009), page numbers (2341–2351).

Habashy, T. M., & Abubakar, A. (2004). A General Framework for Constraint Minimization for the Inversion of Electromagnetic Measurements. *Prog. Electromag. Res.*, Vol. 46, (2004) page numbers (265–312).

Isernia, T., Pascazio, V., & Pierri, R. (2001). On the local minima in a tomographic imaging technique. *IEEE Trans. Geosci. Remote Sens.*, Vol. 39, (2001) page numbers (1596–1607).

Kharkovsky, S., & Zoughi, R. (2007). Microwave and millimeter wave nondestructive testing and evaluation - Overview and recent advances. *IEEE Instrum. Meas. Mag.*, Vol. 10, (2007) page numbers (26–38).

Landweber, L. (1951). An Iteration Formula for Fredholm Integral Equations of the First Kind. *Am. J. Math.*, Vol. 73, (1951) page numbers (615–624).

Lesselier, D., & Bowler, J. (2002). Special section on electromagnetic and ultrasonic nondestructive evaluation. *Inv. Probl.*, Vol. 18, (2002).

Litman, A., Lesselier, D., & Santosa, F. (1998). Reconstruction of a two-dimensional binary obstacle by controlled evolution of a level-set. *Inv. Probl.*, Vol. 14, (1998) page numbers (685–706).

Lobel, P., Blanc-Féraud, L., Pichot, C., & Barlaud, M. (1997). A New Regularization Scheme for Inverse Scattering. Inverse Problems. *Inv. Probl.*, Vol. 13, (1997) page numbers (403–410).

Lobel, P., Kleinman, R. E., Pichot, C., Blanc-Feraud, L., & Barlaud, M. (1996). Conjugate-Gradient Method for Soliving Inverse Scattering with Experimental Data. *IEEE Antennas Propag. Mag.*, Vol. 38, (1996).

Oliveri, G., Lizzi, L., Pastorino, M., & Massa, A. (2012). A Nested Multi-Scaling Inexact-Newton Iterative Approach for Microwave Imaging. *IEEE Trans. Antennas Propag.*, Vol. 60, (2012), page numbers (971–983).

Pastorino, M. (2010). *Microwave imaging.* John Wiley, Hoboken N.J.

Paulsen, K. D., Poplack, S. P., Li, D., Fanning, M. W., & Meaney, P. M. (2000). A clinical prototype for active microwave imaging of the breast. *IEEE Trans. Microw. Theory Tech.*, Vol. 48, (2000), page numbers (1841–1853).

Rekanos, I. T. (2008). Shape Reconstruction of a Perfectly Conducting Scatterer Using Differential Evolution and Particle Swarm Optimization. *IEEE Trans. Geosci. Remote Sens.*, Vol. 46, (2008) page numbers (1967–1974).

Remis, R. F., & van den Berg, P. M. (2000). On the equivalence of the Newton-Kantorovich and distorted Born methods. *Inv. Probl.*, Vol. 16, (2000), page numbers (L1–L4).

Richmond, J. (1965). Scattering by a Dielectric Cylinder of Arbitrary Cross Section Shape. *IEEE Trans. Antennas Propag.*, Vol. 13, (1965) page numbers (334–341).

Rocca, P., Benedetti, M., Donelli, M., Franceschini, D., & Massa, A. (2009). Evolutionary Optimization as Applied to Inverse Scattering Problems. *Inv. Probl.*, Vol. 25, (2009).

Schilz, W., & Schiek, B. (1981). Microwave Systems for Industrial Measurements. *Advances in Electronics and Electron Physics*, pp. (309–381). Elsevier.

Shea, J. D., Kosmas, P., Van Veen, B. D., & Hagness, S. C. (2010). Contrast-Enhanced Microwave Imaging of Breast Tumors: A Computational Study Using 3D Realistic Numerical Phantoms. *Inv. Probl.*, Vol. 26, (2010).

van den Berg, P. M., & Abubakar, A. (2001). Contrast Source Inversion Method: State of Art. *Prog. Electromag. Res.*, Vol. 34, (2001) page numbers (189–218).

Vogel, C. R. (2002). *Computational methods for inverse problems.* Society for Industrial and Applied Mathematics, Philadelphia.

Zhang, Z. Q., & Liu, Q. H. (2004). Three-Dimensional Nonlinear Image Reconstruction for Microwave Biomedical Imaging. *IEEE Trans. Biomed. Eng.*, Vol. 51, (2004), page numbers (544–548).

Zhou, H., Takenaka, T., Johnson, J., & Tanaka, T. (2009). A Breast Imaging Model Using Microwaves and a Time Domain Three Dimensional Reconstruction Method. *Prog. Electromag. Res.*, Vol. 93, (2009) page numbers (57–70).

Zoughi, R. (2000). *Microwave Non-Destructive Testing and Evaluation.* Kluwer Academic Publishers, Dordrecht.

Chapter 5

Relevance of Dielectric Properties in Microwave Assisted Processes

Anna Angela Barba and Matteo d'Amore

Additional information is available at the end of the chapter

1. Introduction

Microwaves are electromagnetic radiation with wavelength ranging from 1 mm to 1 m in free space with a frequency from 300 GHz to 300 MHz, respectively. International agreements regulate the use of the different parts of the spectrum; the frequencies 915 MHz and 2.45 GHz are the most common among those dedicated to power applications for industrial, scientific and medical purposes (Metaxas & Meredith, 1983).

Although microwaves have been firstly adopted for communications scope, an increasing attention to microwave heating applications has been gained since World War II (Meredith, 1998; Chan & Reader, 2002). Reasons for this growing interest can be found in the peculiar mechanism for energy transfer: during microwave heating, energy is delivered directly to materials through molecular interactions with electromagnetic field via conversion of electrical field energy into thermal energy. This can allow unique benefits, such as high efficiency of energy conversion and shorter processing times, thus reductions in manufacturing costs thanks to energy saving. Moreover, other effects have been pointed out, such as the possibility to induce new structural properties to irradiated materials (development of new materials) and to apply novel strategies in chemical syntheses (green techniques).

Crucial parameters in microwave heating are the dielectric properties of matter; they express the energy coupling of a material with electromagnetic microwave field and, thus, the heating feasibility (Metaxas & Meredith, 1983; Schubert & Regier 1995; Tang et al., 2002). On the basis of dielectric properties, microwave devices (applicators) can be adopted in heating operations and optimized working protocols can be used.

This chapter is divided into four sections dealing with:

i.   fundamentals of microwave heating and relevance of dielectric properties of materials;

ii.   different techniques used in dielectric properties measurements of materials (test fixtures characteristics, technique applicability, advantages and disadvantages);

iii. application of the open-ended coaxial-probe method in dielectric properties measurements of food, pharmaceutical ingredients, living materials, to understand specific heating phenomenology and, thus, to optimize thermal treatments / to define safety limits of exposition;

iv.   basics of heat and mass transfer modeling in microwave assisted processes.

2. Microwave heating fundamentals

Peculiarity of microwave heating is the energy transfer. In conventional heating processes, energy is transferred to material by convection, conduction and radiation phenomena promoted by thermal gradients and through the materials external surface. Differently microwave energy is delivered directly to materials through molecular interactions (loss mechanisms) with electromagnetic field via conversion of electromagnetic energy into thermal energy. Whereas loss mechanisms occur, a high rate of heating and a high efficiency of energy conversion are expected. The high heating rate represents the key-feature of microwaves heating, because this makes possible to accomplish in short times (seconds or minutes) what would take minutes, or even hours, to be done with conventional heating. This depends upon slowness of heat delivery rate from the material surface to the core as determined by the differential in temperature from a hot outside to a cool inside. In contrast, use of microwave energy can produce, under some conditions, a bulk heating with the electromagnetic field interacting with the material as a whole. With reference to energy saving, thermal treatments performed by microwave heating can be seen as intensified operations[1].

The ability of a material to interact with electromagnetic energy is related to the material's complex permittivity (dielectric properties or susceptibility). This property, in any homogenous, isotropic, and linear dielectric material is characterized by a frequency-depending absolute complex permittivity usually indicated with the Greek symbol ε:

$$\varepsilon_{abs}(\omega) = \varepsilon_0 \ \varepsilon(\omega) = \varepsilon_0 \left[\varepsilon'(\omega) - i\varepsilon''(\omega) \right] \tag{1}$$

where ε_0 is the vacuum permittivity ($\varepsilon_0 = 8.85 \ 10^{-12}$ F/m) and ω is the angular frequency ($2\pi f$, f frequency, Hz). In scientific literature, complex permittivity is diffusely reported as a relative complex number $\varepsilon = \varepsilon_{abs} / \varepsilon_0 = \varepsilon' - i\varepsilon''$ in which the real part, ε', is named dielectric constant and the imaginary part, ε'', is known as loss factor. The dielectric constant is a measure of how much energy from an external electric field is stored in the material; the loss

---

[1] Process intensification is a current approach in the development of equipment and methods to achieve process miniaturization, reduction in capital cost, improved energy efficiency, and, often, product quality. Additional benefits of process intensification include improved intrinsic safety, simpler scale-up procedures. The philosophy of process intensification has been traditionally characterized by four words: smaller, cheaper, safer, slicker (Coulson & Richardson's, 2002; Stankiewicz, A. & Moulijn J. 2004).

factor accounts for the loss energy dissipative mechanisms in the material[2]. Therefore, a material with a high loss factor is easily heated by microwave. On the other hand, if a material has a very low ε'' is transparent to microwave effect. Power dissipation (Q_g) is given by the common form of the average power loss density (power dissipation per unit volume, W/m³) drawn from the Poynting's theorem (Metaxas & Meredith, 1983):

$$Q_g = \frac{1}{2}\,\omega\varepsilon_0\varepsilon''\,|E|^2 \qquad (2)$$

where E is the electrical field strength [V/m]. Bulk heating is achieved when penetration depth (D_p), defined as the distance from the material surface at which the power drops to e^{-1} of its initial value, is of the same order of magnitude of materials dimensions. Assuming electromagnetic field as a plane wave that travels along one axis, penetration depth is calculated as following:

$$D_p = \frac{c}{2\sqrt{2}\pi f\sqrt{\varepsilon'}\left[\sqrt{1+\tan^2\delta}-1\right]^{\frac{1}{2}}} \qquad (3)$$

where c is the light velocity in free space ($3\ 10^8$ m/s) and $\tan\delta$ is the loss tangent. Under some conditions ($(\varepsilon''/\varepsilon') \ll 1$, i.e. small $\tan\delta$) the penetration depth can be calculated by:

$$D_p = \frac{c}{2\pi f}\frac{\sqrt{\varepsilon'}}{\varepsilon''} \qquad (4)$$

When bulk heating is not achievable, a temperature levelling effect can occur in thick layer depending on materials thermal diffusivity that can drive the heat distribution within the whole bulk.

Under a physical point of view, interactions between materials and electromagnetic energy are inherent in the ability of the electric field to polarize the material charges and in the impossibility of this polarization to follow the rapid changes of the oscillating electric field (dielectric dissipative mechanisms). In presence of an external electric field, different kinds of polarization mechanisms are possible: the electronic polarization, caused by modification of electrons position around the nucleus; the atomic polarization, related to positional shifts of nucleus due to the non-uniform distribution of charges within the molecule; the orientation polarization (dipoles rotation) due to the reorientation of the permanent dipoles under the influence of the electric field; the spatial charge polarization observed in materials containing free electrons confined on surface (Maxwell-Wagner effect) (Metaxas and Meredith, 1983). Depending on frequency, one or two mechanisms dominate over the others. In particular, among the dielectric mechanisms of energy dissipation above outlined,

---

[2] The loss tangent ($\tan\delta = \varepsilon''/\varepsilon'$) is frequently used in dielectric heating literature providing indications of how the material can be penetrated by an electric field and how it dissipates the energy in heat.

the dipoles rotation is the dominant polarization mechanism in irradiating materials rich in water (such as biological tissues, foods, mixtures based on water or polar solvents) in the microwave electromagnetic spectrum region (industrial high frequency heating $10^7<f[Hz]<10^9$). In the same region also ionic dissipative phenomena (Joule's loss effect) may occur if ionic species are present, and can be lossy. The atomic and the electronic polarization mechanisms are relatively weak, and usually constant over the microwave region (Fig. 1.).

Being in dielectric measurements difficult to separate conduction losses from those due polarization, the overall dissipative feature (loss factor) of a material can be expressed by the following equation:

$$\varepsilon''_{measured}(\omega) = \varepsilon''_{ep}(\omega) + \varepsilon''_{ap}(\omega) + \varepsilon''_{dp}(\omega) + \varepsilon''_{int\,erfp}(\omega) + \frac{\sigma}{\varepsilon_0 \cdot \omega} = \varepsilon''(\omega) + \frac{\sigma}{\varepsilon_0 \cdot \omega} \tag{5}$$

where the subscript *ep, ap, dp* and *intefp* refer to electronic, atomization dipolar and interfacial polarization mechanisms, respectively; and σ is the conductivity of the medium.

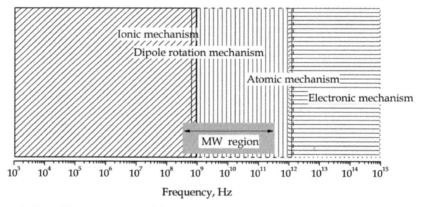

Figure 1. Typical frequency-regions of the loss mechanisms.

Microwave heating processes are currently applied in many fields: from food industry, including packaging (Tang et al., 2002; Schubert & Regier, 2005) to materials processing (polymers, wood, ceramics and composites) (Zhou et al., 2003); from minerals treatments (Al-Harahsheh & Kingman, 2004) and environmental remediation processes (soil remediation, toxic waste inertization) (Kulkarni et al., 2008; Remya & Lin, 2011; Barba et al., 2012) to pharmaceutical emerging technologies (McMinn et al., 2005; Auriemma et al., 2011). Key of all processes above is the energy transfer that, as first discussed, is based on the ability of a material to store and to dissipate electromagnetic energy. Knowledge of the dielectric properties appears fundamental for heating treatments because they have a crucial role in designing (or choosing) of microwave devices (applicators) and on setting operative parameters (time of exposure, power). Moreover, as dielectric properties can be affected by many factors including frequency of microwaves, temperature, chemical composition of the materials (abundance of water, salt content and other constituents) (Kraszewski, 1996; Chan & Reader, 2002; Tang et al., 2002), microwave heating requires to be appropriately addressed on the basis of dielectric behaviour studies.

3. Dielectric properties measurements

Studies on materials dielectric behaviour are performed by measurements using dedicated instruments and methods. In particular, several techniques have been developed to measure the dielectric properties of materials in the microwave region. They are fundamentally based on the use of vector network analysers (VNA). These instruments, today managed via personal computer, are able to perform measurements of reflection/transmission by opportune test-fixtures. Starting from acquired signals and implemented electromagnetic model, dielectric properties are calculated via software. It is possible to operate with different kinds of test-fixtures developed for various classes of material such as fluids, powders, films. Then, the same general consideration in choosing technique/fixture must be adopted (i.e. destructive *vs* non-destructive analysis; isotropic *vs* non-isotropic samples, permeability or permittivity as relevant property)[3].

In the following sub-sections different measuring methods are briefly presented emphasizing applicability, advantages and disadvantages. In all methods calibration procedures must be performed to reduce systematic errors.

3.1. Open-end coaxial line method

A VNA is connected via coaxial cable to an open-end coaxial probe. To perform measurements the technique requires to press the coaxial probe against the sample material (Fig. 2). The microwave signal launched by the VNA is reflected by the sample and basing on the reflected waves, both dielectric constant and loss factor are calculated. The main advantages of this technique are that it is easy to use, is non-destructive, is very fast and has a reasonable accuracy (± 5%). It is well suited for samples with flat surface (best for liquid or semisolids) with homogeneous features and with not so high dielectric constant or too low loss factor. It requires little or no sample preparation. Since the coaxial probe is produced with rugged structure and materials, the technique also allows to perform measurements in a wide –40 to +200° C temperature range and for corrosive or abrasive materials. On the other side, by the open-end coaxial probe technique, samples with rough surfaces, very thin or anisotropic structures, or at high temperature or with permeability features cannot be investigated. Moreover, use of an open-ended coaxial probe device requires strict operative conditions: cable stability (no flexing of the coaxial cable must occur); absence of air gap between probe and samples surface (absence of bubbles in liquid samples and flat surface in

---

[3] Technical brochures of VNA manufacturers are useful in fixtures choices, uses, instruments better performances. As an example see: *http://cp.literature.agilent.com/litweb/pdf/5989-2589EN.pdf* (Agilent Technologies company).

solids and semi-solid samples); use of appropriate sample thickness (measurements must be performed on a "semi-infinite" sample).

Globally, this technique is most commonly used by the research community, especially in microwave food processing, for the features above presented and for the affordable cost of the open-end coaxial probe fixture.

3.2. Transmission line method

Transmission line method requires a VNA connected with a coaxial or rectangular wave guide where the sample under investigation completely fills the wave guide cross section. Dielectric measurements are based on the change of impedance and propagation characteristics of microwave signal launched by the VNA. This method can give accurate measurements and provides both permittivity and permeability measurements, but requires a careful sample preparation (no air gap at fixture walls is tolerated). The method is more expensive then the open-end coaxial probe system (referred to the same range of frequencies), is not easy to use and is time-consuming. Finally, samples with low loss cannot be investigated.

3.3. Free space method

When the contact between fixtures and sample under investigation is not possible (corrosive materials, high temperature) dielectric and also permeability measurements can be performed using the free space method. This method is applied using a VNA connected to two antennas that "see" the interposed samples. The free space method, thus, has the potentiality to perform on-line dielectric behaviour studies of materials under treatments, as an example, in furnaces. The method is not destructive but requires homogeneous, large and flat samples (films, slabs).

3.4. Resonance cavity method

VNA is connected with a resonant cavity where the sample under investigation is placed. This method is based on the shift of the center resonant frequency and the alteration of the cavity quality factor due to the presence of the material in the cavity. It is suitable to investigate very small samples and low loss materials (it also provides permeability measurements) that cannot be investigated with the methods above reported, globally, giving an accuracy of ± 2-3%. This method has complex procedure, requires shaped samples (destructive analysis) and the fixture is more expansive than the open-end coaxial probe.

4. Applications of the open-ended coaxial-probe method

In the following sub-sections examples of dielectric proprieties measurements are presented and discussed. In particular, dielectric spectroscopies of different materials (fruits and vegetables, pharmaceutical mixtures, biological tissues) are performed with the aim of investigating their behaviour when exposed, at given conditions, to microwave irradiation.

Measurements are carried out to respond to two different purposes: to develop *ad-hoc* working protocols (in unit operations based on microwave heating such as blanching, cooking, drying) and to evaluate electromagnetic risks for living tissues if exposed to electromagnetic phenomena (specific absorption rate, SAR, estimations).

Being the materials under investigation basically aqueous, notes on water and saline solutions dielectric behaviour are reported in the first sub-section. In particular, the role of frequency temperature and salt content is emphasized.

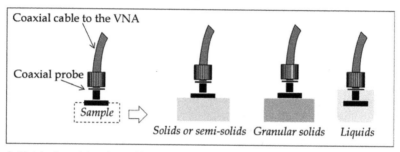

Figure 2. Sketch of coaxial probe position *vs* material under test during dielectric properties measurements.

Note that all the investigated materials are subjected to dielectric spectroscopy in the frequency range 200 MHz – 6 GHz using an open-ended coaxial probe device (Agilent Technologies mod. 85070D) connected to a VNA (Agilent Technologies, mod. ES 8753). Due to the similar nature of the investigated materials, the instrument calibration procedure was performed using the same standards: air; distilled water at 25°C and short circuit block. In Fig. 2. a sketch of the open-end coaxial probe coupled with different samples under investigation is reported.

Even if calibration protocol ensures the removal of systematic errors, additional sources of errors can affect the accuracy of this measurement method. In particular, as reported above, the use of open-ended coaxial probe device requires the operative conditions of: cable stability, absence of air gap between probe and samples surfaces, appropriate sample thickness. A rough estimation of this latter parameter can be calculated by:

$$\text{Sample thickness} = \frac{1}{0.05 \cdot \sqrt{|\varepsilon|}} \ [\text{cm}] \tag{6}$$

where $|\varepsilon|$ is the module of the expected complex permittivity. Higher loss features will require lower sample thicknesses.

4.1. Dielectric properties of water and saline solutions

Water is a medium constituted by natural dipoles which can become polarized under application of electric fields orienting the molecules. Dielectric spectroscopy of distilled water at different temperature is reported in Fig. 3. (Mudgett, 1986; Barba & d'Amore 2008). Al low frequencies ("static" region), water dipoles have enough time to follow variations of the applied field and dielectric constants are at their maximum value (maximum energy of the external source is stored in the material) whereas loss factors are characterized by very low values, approaching zero. Similarly, at high frequency ("optical" region, towards the region of infrared and visible radiations) water dipoles fail to follow the rapid oscillations of the external applied fields so a reduction in polarization effect occurs. This is manifested by the low values of the dielectric constants and of the loss factors low values, i.e. about 5 and zero, respectively. In the intermediate region, known as dispersion" or "relaxation", the dielectric constants decrease as the frequency increases, passing from the static value to the optical one. In this region the dipoles pass from one situation in which are apparently undisturbed because they fully restore their original position during field reversals, to a situation in which they are really undisturbed because the stresses are continuing at frequencies too high to allow the orientation and then relaxation. In intermediate conditions, instead, microwaves frequencies are able to induce phenomena of orientation and consequently of relaxation. In particular, in this spectral region, water loss factor shows a maximum, localized in correspondence of a critical frequency, f_c, which corresponds to a critical wavelength, λ_c. For water at 25°C, the critical frequency is about 18 GHz and the critical wavelength corresponds to 0.017 m. Orientation and relaxation phenomena are thus strongly a function of microwaves frequency. A key parameter used to define orientation effect is the relaxation time τ. It represents the time required to relax the dipolar orientation produced by a static electric field, starting from the removal of the field up to a percentage equal to $1 / e = 0,37$ of the initial value. The relaxation time is related to the frequency critical wavelength by:

$$\tau = \frac{1}{\omega_c} = \frac{1}{2\pi \cdot f_c} = \frac{\lambda_c}{2\pi \cdot c} \tag{7}$$

where ω_c is the angular critical frequency ($\omega_c = 2\pi f_c$) and c is light velocity in free space ($c = 3 \cdot 10^8$ m s⁻¹). Relaxation time is a decreasing function of temperature (the critical frequency shows a shift towards high frequencies as temperature increases). Temperature affects both water viscosity and bulk Brownian motions whose variations increase the mobility of the water molecules. By this way water dipoles can easily escape to the orienting action of the applied fields, thus leading to a decrease of the dissipative effect (see also reduction of dielectric properties values). Relaxation time dependence from temperature can be evaluated according to a law Arrhenius type equation (Tang et al., 2002):

$$\tau(T) = \frac{A}{T} \exp\left(\frac{B}{T}\right) \tag{8}$$

where A and B are parameters and T is the absolute temperature [K].

An advantage of the decrease of the dissipative character at high temperature is the achievement of the so called "leveling effect" which, depending from thermal properties of materials, can drive a heat redistribution. This phenomena may play an important role in the heating processes: if dielectric losses cause a local overheating (hot-spot), a corresponding decrease of dissipative features occurs and, in turn, the growing importance of conductive and convective transport mechanisms contribute to a better heat distribution.

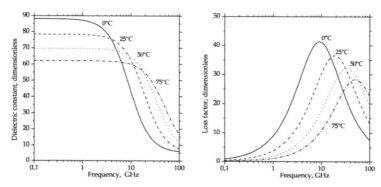

Figure 3. Dielectric spectroscopy of water (dielectric constant on left, loss factor on right) at different temperatures (Barba & d'Amore, 2008, redraw from Mudgett, 1986).

It is important to note that even if dielectric relaxation of liquid water has been extensively studied, the dielectric properties of water in bulk are not of any major significance in industrial microwave heating applications since the relaxation of bounded water is far more important for the majority of applications (Metaxas & Meredith, 1983). Nevertheless, knowledge of parameters that affect dielectric dissipation phenomena (see also dielectric properties) are basilar to understand the relative predominance of the different loss mechanisms at allocated frequencies for industrial, scientific and medical purposes.

Indeed, in many cases it is of interest to treat materials in which liquid water may be present at free and bounded state. Because of the greater difficulty with which the dipoles oscillate due to interactions with the molecules of materials, bounded water presents longer relaxation times (critical frequency is shifted towards lower frequencies thus maximum values of loss factors are achieved under 18 GHz). Fig. 4. depicts, qualitatively, the classical relationship between loss factor and moisture content of a material. Distinct inflexion points in the profile demarcate the transition between the changing states of water in the material. Generally, at low moisture content water primarily exists in bounded form, thus possessing a limited mobility in the presence of electromagnetic waves. As moisture content in material increases, a critical level (or critical moisture content, M_c) is attained. Physically, this means that all the available binding sites for water molecules becomes saturated. Further additions of water beyond this critical level result in an increase in population of free water molecules and consequently in an increase of dielectric losses. In many cases, dielectric properties measurements are performed to determine the M_c of solid mixtures.

The effect of temperature on the bounded water shows an opposite trend with respect to the one exhibited from free water. In fact, the dielectric properties increase with temperature, since the bounded dipoles become free from links and more available to the orientation effects induced by applied electromagnetic fields. This physical behavior is called "thermal runaway" and, briefly, can be described as a progressive increase of loss features of heated material. Thermal runaway, if not properly taken into account or controlled, can have deleterious effects because it promotes overheating phenomena.

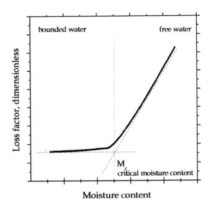

Figure 4. Qualitative relationship between moisture content and loss factor. Mc refers to the critical material moisture content.

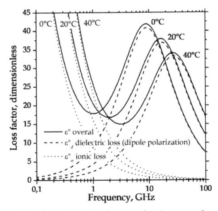

Figure 5. Ionic conduction and dipole polarization loss mechanisms as a function of temperature in NaCl water solution (Barba & d'Amore, 2008, redraw from Mudgett, 1986).

Presence of ionic species is another relevant aspect to consider if aqueous materials are exposed to microwave fields. Presence of ions in liquid or in wet solid bulks, in general, involves decreases of dielectric constant values and increases of loss features. Decrements of the energy-storage ability are due to subtractions of water dipoles, because solvation effects occur (less dipoles can be polarized); loss factor values increments are due to the enhance of ionic dissipative mechanism, especially at low frequencies. Both described effects increase with increasing temperature. Effects of ionic species in water bulk, at different temperatures, are investigated by NaCl solutions dielectric measurements (Mudgett, 1986). In Fig. 5. loss factor evolution vs frequency of NaCl water solution 0.5 M is shown to emphasize the two main loss mechanisms occurring (dipole polarization, ionic dissipation). One can easily verify that, for a given frequency and for a given concentration, the loss factor appears to be decreasing with temperature until the effect of the dipoles dominates and, then, increasing, when the effect of the charges prevails. For example, to a solution 0.1 M at 2.45 GHz, the loss factor ranges from 25 to 20 as the temperature increases from 0° C to 100°C, with a minimum of about 17 at a temperature of about 40°C (Barba & d'Amore, 2008). Temperature also affects viscosity medium properties increasing the mobility of charges.

4.2. Dielectric properties of fruits and vegetables

Thermal treatments are the most diffused processes applied in agro-foods manufacturing to obtain products with prolonged shelf-life and suitable to eating for enhanced digestibleness, flavours and rheological properties. Sterilization, pasteurization, pre-cooking and cooking, tempering of frozen foods and drying are the unit operations commonly applied (Metaxas, 1983; Schubert & Regier, 2005). Moreover, studies have been focused on the possibility of using electromagnetic energy to disinfest fruit harvests from insect pests (Wang et al., 2003) or to detect, by nondestructive investigation, the amount of moisture in materials by sensing the dielectric properties of the material (McKeown et al., 2011).

Despite the enormous industrial interest in food processing / characterization, the potentialities of microwave heating are not fully developed yet at industrial scale. Reasons of this delay include lacks of basic information on the dielectric properties of food and their relationships to microwave heating / material interaction characteristics and on equipment costs in a technology that has not been proven thoroughly reliable in large scale or long term uses (Ohlsson, 2000; Tang, Hao & Lau, 2002; Schubert & Regier 2005). However, world researches are currently developed on these subjects as proven by the hundreds of papers published each year on microwave / food / dielectric properties measurements topics and plants development.

Agro-foods show high interactions with the principal wave-frequencies used in industrial heating (915 MHz and 2.45 GHz), so they are raw materials suitable to be processed by microwave energy. In Figs. 6. and 7. dielectric constant and loss factor measurements of agro-foods (some kinds of fruits and vegetables) are reported. Accurate measurements have been carried out in fruits dielectric properties measurements due to the softer structure (pulp) that is optimal for a positioning under the coaxial probe. Conversely, in vegetables matrices measurements, air gaps, due to rough and fibrous surfaces, in some cases, strongly affected the performed measurements.

Both fruits and vegetables dielectric spectroscopies show that, due to moisture and soluble solids contents of agro-foods, their dielectric behavior are dominated by dipolar polarization and ionic loss mechanisms. Indeed, dielectric constant profiles are characterized by typical monotonic trend and depressed values observed for of saline solution; loss factor signal shapes, especially at low frequency, highlight the predominance of an ionic dissipation mechanism. Soluble solids contents of agro-foods (measured by °Brix) can be referred to sugar, starch and mineral components. Hydration of sugar and starch molecules reduces dielectric constants; presence of minerals affects loss factors increasing the dissipative features. Moreover, relaxation time shifts to low frequencies.

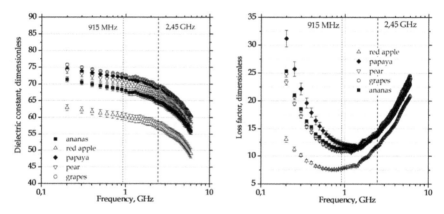

Figure 6. Dielectric spectroscopy of fruits (dielectric constant on left, loss factor on right; for fruits properties see Table 1.).

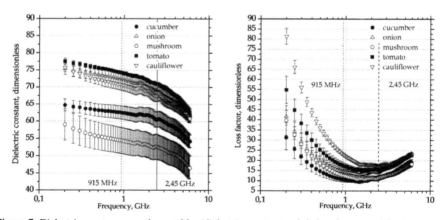

Figure 7. Dielectric spectroscopy of vegetables (dielectric constant on left, loss factor on right; for vegetables properties see Table 2.).

Fruit	Moisture content % w/w	°Brix %	Dielectric constant (2.45 GHz) ε' ± SD	Loss factor (2.45 GHz)ε"± SD	Temperature °C
Ananas	86.00± 0.47	16	64.89 ± 0.99	13.98 ± 0.13	20
Banana	67.87±0.93	12.5	59.28 ± 0.58	17.72 ± 0.31	22
Black grapes	83.49± 0.22	16	66.11 ± 1.88	15.57 ± 0.91	20
Black plum	82.82±20	15	61.36 ± 5.53	14.38 ± 1.32	20
Cherry	71.22±5.61	12	70.08 ± 1.47	18.87 ± 0.36	22
Grapes	82.66±1.17	12.5	69.45 ± 0.09	14.35 ± 0.08	20
Green apple	82.37± 0.42	9	35.99 ± 0.86	7.06 ± 0.04	21
Kiwi	79.48±2.32	16	61.22 ± 0.12	15.72 ± 0.02	22
Lime	89.34± 0.34	7	59.20 ± 11.47	13.73 ± 3.10	20
Orange	83.54± 0.20	11	67.40 ± 0.17	14.37 ± 0.28	20
Papaya	81.62±3.47	12.5	69.00 ± 0.25	13.86 ± 0.28	22
Passion fruit	59.11± 5.14	13	58.13 ± 1.61	13.68 ± 1.61	22
Peach	87.73± 0.51	12.5	61.42 ± 14.07	11.07 ± 2.79	22
Pear	83.01±1.42	14	67.51 ± 1.05	14.16 ± 0.22	20
Red apple	87.02± 0.72	10	57.56 ± 1.18	11.72 ± 0.22	22
Strawberry	84.27±6.03	9	68.55 ± 2.05	12.40 ± 0.28	22

Table 1. Water content, °Brix, dielectric properties (at 2.45 GHz) of fruits.

Vegetable	Moisture content % w/w	°Brix %	Dielectric constant (2.45 GHz) ε' ± SD	Loss factor (2.45 GHz)ε"± SD	Temperature °C
Cauliflower	84.90± 0.34	-	67.46 ± 1.66	17.42 ± 0.03	22
Cucumber	94.67± 0.09	3	60.49 ± 2.18	11.53 ± 0.21	22
Mushroom	91.90± 1.55	-	52.69 ± 3.63	11.50 ± 1.17	22
Onion	91.96± 1.60	-	69.06 ± 1.69	14.32 ± 0.43	22
Pepper	91.6± 4.30	7	59.10 ± 12.68	11.04 ± 2.48	21
Potato	69.50± 0.30	-	61.13 ± 5.41	16.63 ± 1.58	22
Ravanello	96.15± 1.95	-	68.88 ± 9.32	12.59 ± 1.94	22
Tomato	91.14± 0.22	9	70.87 ± 0.30	15.70 ± 0.50	21
Zucchini	95.50± 0.73	-	63.98 ± 3.67	15.02 ± 1.12	22

Table 2. Water content, °Brix, dielectric properties (at 2.45 GHz) of vegetables.

Temperature effects on dielectric properties of fruits matrices (banana: moisture content ~68%, °Brix=12.5; ananas: moisture content ~86%, °Brix=16) are shown in Figs. 8. and 9. As reported above, temperature affects dipolar dissipative mechanism because increases the molecular disorder and, thus, the reduction of dielectric properties values. Consequently, penetration depth increases (Fig. 9.). This latter behavior can assume a relevant role if bulk heating are desired in large samples where an uniform temperature is not achieved due to the low penetration depth. Due to the influence of many parameters, it is difficult to develop a general predictive equation to accurately describe dielectric properties of foods. Eq. 9 summarizes the main factors that must be taken in account:

$$\varepsilon\left(f,T,M,\underline{\phi}\right)=\varepsilon'\left(f,T,M,\underline{\phi}\right)-ie''\left(f,T,M,\underline{\phi}\right) \tag{9}$$

where T is the temperature, M the moisture content, $\underline{\phi}$ is the vector of properties such as conductivity, heat capacity, molecular structure, density, size, crystallinity and many other physical properties of materials.

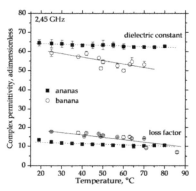

Figure 8. Dielectric properties of ananas and banana fruits vs temperature.

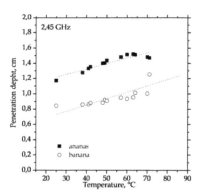

Figure 9. Penetration depths vs temperature calculated by eq. 4.

Under some conditions (i.e. given frequency for heating applications, constancy of materials properties etc...), empirical correlation are developed. In Sipahioglu & Barringer, 2003, empirical polynomial equations were proposed to estimate dielectric properties of agro foods:

vegetables:

$$\varepsilon' = -243.6 + 1.342 \cdot T + 4.593 \cdot M - 426.9 \cdot A + 376.5 \cdot A^2 - 0.01415 \cdot M \cdot T - 0.3151316.2 \cdot A \cdot T \quad (10)$$

$$\varepsilon'' = -100.02 - 0.1611 \cdot T + 0.001415 \cdot T^2 + 2.49 \cdot M - 378.9 \cdot A + 316.2 \cdot A^2 \quad (11)$$

fruits:

$$\varepsilon' = 22.12 + 0.2379 \cdot T + 0.5532 \cdot M - 0.0005134 \cdot T^2 - 0.003866 \cdot M \cdot T \quad (12)$$

$$\varepsilon'' = 33.41 - 0.4415 \cdot T + 0.001400 \cdot T^2 - 0.1746 \cdot M + 1.438 \cdot A + 0.001578 \cdot M \cdot T + 0.2289 \cdot A \cdot T \quad (13)$$

where T is the temperature, (°C), M and A are the moisture (% w/w dry basis) and the ash (%w/w dry basis) contents, respectively.

4.3. Dielectric properties of pharmaceutical aqueous mixtures

Interest in applications of microwave power (i.e. dielectric drying, curing) and related techniques for non-destructive, on-line moisture determination of wet powders and granules in pharmaceutical industry is relatively recent and not widespread. A non-consolidated know-how makes prudent industrial investments whereas, under a technical point of view, the need to have a uniform distribution of heat to avoid local overheating of the material, and, moreover, the possibility of a temperature uncontrolled increase are the main limiting factors (Heng et al., 2010; Barba et al., 2010a). In the last twenty years, mainly in Academic researches and in a few large-scale plants experiences, microwaves benefits have begun to capture attention. In particular, microwaves have been applied in modulating drug and excipient properties via specific material microwave interactions (Wong et al., 2002; Nurjaya & Wong, 2005; Auriemma et al., 2011); tested in drying operations of powder and granules (Hegedus & Pintye-Hodi, 2007; Loh et al., 2008); used in single-pot devices to carry out mixing, granulation and drying of pharmaceuticals in one vessel (Kelen et al., 2005; McMinn et al., 2005; Hegedus & Pintye-Hodi, 2007). Moreover, dielectric properties measurements of pharmaceutical powders wet mixtures are used to control moisture / solvent residues (Gradinarsky et al., 2006).

Once again, as in food industries, knowledge of the mechanisms and of the intensity of the interaction fields electromagnetic / pharmaceuticals is a fundamental step for any consolidation in equipment development and new processing methods.

In Fig. 10 dielectric properties of several common excipients (dry powders, some properties in Table 3.), at 2.45 GHz and room conditions, are reported (PVP, PolyVinyl-Pyrrolidone; CMC, Carboxy-Metil-Cellulose).

As can be easily seen, the excipients show very low dissipative features and, in turn, penetration depths assume high values (Fig. 11.). The effect of water content on dielectric properties of lactose and CMC distilled water mixtures is reported in Fig. 12 (at 2.45 GHz, room conditions).

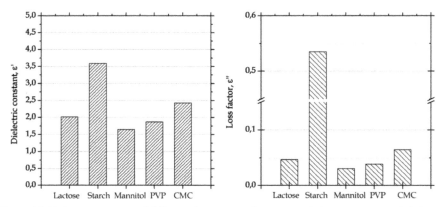

Figure 10. Dielectric properties of different pharmaceutical excipients (dielectric constant on left, loss factor on right) at 2.45 GHz, room temperature.

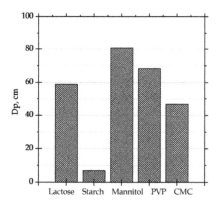

Figure 11. Penetration depth in different pharmaceutical excipients (water content at storage condition: lactose 0.5%; starch 9.4%; mannitol 0.19%; PVP 3.5%; CMC 6.5%) at 2.45 GHz, room temperature.

Excipient	Water content % w/w	D_p cm	Dielectric properties (2.45 GHz) $\varepsilon' - \varepsilon''$	Conductivity* µS/cm	Temperature °C
CMC	50%	0.50	35.6 – 23.2	4060.00	25
Lactose	50%	1.15	50.41-12.11	12.75	25
Mannitol	50%	1.43	56.9 - 10.3	6.03	25
PVP	50%	0.81	29.5 - 13.0	79.77	25
Starch	50%	1.42	53.9 -10.1	16.48	25
*conductivity of excipient/distilled water at 0.02 g/L – distilled water conductivity: ~ 5.00 µS/cm					

Table 3. Excipient/water mixtures (50%w/w) data: dielectric properties, penetration depths, conductivities of diluted solutions.

In both kinds of mixtures, dielectric constant values show a monotone increasing trend as a function of water content, even if higher values are achieved in lactose mixtures rather than in CMC mixtures at equal water contents. Moreover, different critical moisture points are exhibited (Mc lactose~ 20%; Mc CMC~40%). This is due to the saline form of CMC powders (sodium salt) that gives ionic species in water medium and, indeed, the contribute of ionic dissipation mechanisms is observed in loss factor profiles. In particular, the effect of the ionic dissipation mechanism is clearly highlighted in the mixtures dielectric spectroscopies. In Figs. 13. and 14. dielectric properties of lactose and CMC water mixtures (50% w/w) as function of frequency and at different temperature are reported. Dielectric constant profiles show a decreasing trend with an opposite behavior at high temperatures. In lactose mixtures, a high temperature depresses the dielectric constant values because of molecular disorder, and in CMC mixtures a high temperature slightly increases the ionic species mobility. In loss factor profiles of CMC mixtures, the ionic dissipation mechanisms become very relevant at low frequency and increase with temperature because of an increased ionic conductivity as a result of viscosity reduction at high temperature. However, in loss factor profiles of lactose mixtures, temperature water behavior is kept unvaried.

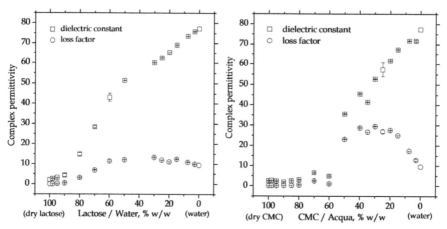

Figure 12. Dielectric properties of lactose / water mixtures (on the left) and of CMC / water mixtures (on the right) (at 2.45 GHz, room conditions).

4.4. Dielectric properties of biological tissues

Safe use of electromagnetic (EM) fields for medical purposes and, on the other hand, for the developing of protection systems from electronic devices leakages caused by industrial and domestic appliances, mobile antennas and so on, require to clarify the interaction between non-ionizing electromagnetic fields (EM) and living systems. Crucial point of investigations on any deleterious impact (overheated tissues) or induced benefit (diathermy, rewarming from hypothermia and cancer treatments) is to quantify the specific absorption rate[4] (SAR) of the biological systems, i.e. to study their dielectric behavior during exposure to electromagnetic fields (Metaxas & Meredith, 1983; Adair & Petersen, 2002). Amount and distribution of the energy adsorbed in a biological system exposed to EM field is related to the internal electric and magnetic fields. As a wave penetrates in a biologic system, the electric field interacts at the various tissue interfaces resulting in a complex distribution of the local fields. These internal fields are related to a number of parameters including frequency, dielectric properties of the tissues, geometry and relative position from EM source. Currently, tools used to evaluate the interactions are either experimental measurements by thermographic or dielectric spectrometry methods (Guy, 1971, Gabriel C., 1996; Gabriel S., 1996a; Gabriel S., 1996b), and numerical simulation procedures (Foster & Adair, 2004).

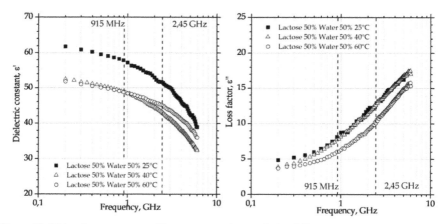

Figure 13. Dielectric spectroscopy of lactose water mixtures (1:1) at different temperatures.

Once more, dielectric properties measurements constitute a necessary step to estimate interaction and thus heating phenomena between fields and materials. In many experimental investigations, dielectric properties measurements of biological tissue are carried out in vivo on accessible parts of the body, ex vivo on fresh excised tissues and in tissue-equivalent materials (Barba et al., 2010b).

In Fig. 15. dielectric properties of bovine fresh excised tissues (liver and brain), measured by the open-end coaxial line method (snapshot of measurements in Fig. 16.), are shown. In general, due to the high moisture content (investigated tissues: liver 68-70%, brain 82-94%) and the saline species, biological tissues are highly receptive to microwave radiation.

---

[4] *SAR* of biological systems is defined as the time derivate of incremental energy (*dQ*) adsorbed by (dissipated in) an

incremental mass (*dm*) contained in a volume element (*dV*) with a given density ρ: $\mathrm{SAR} = \dfrac{d}{dt}\left(\dfrac{dQ}{dm}\right) = \dfrac{d}{dt}\left(\dfrac{dQ}{\rho dV}\right)$.

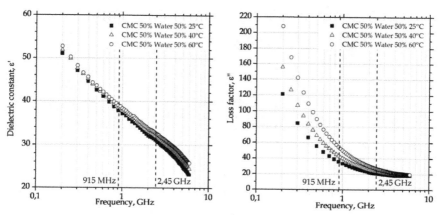

Figure 14. Dielectric spectroscopy of CMC water mixtures (1:1) at different temperatures.

In absence of phase changes, energy rate dissipation calculations could be performed starting from the classic equation of the power dissipation (eq. 2) using the values of measured dielectric properties and appropriately referring the power dissipation term (in eq. 14, Q is the dissipated energy amount [KJ]):

$$SAR = \frac{d}{dt}\left(\frac{dQ}{\rho dV}\right) = \frac{\dot{Q}_g}{\rho} = Cp \cdot \frac{dT}{dt} \qquad (14)$$

It is worth noting that the application of eq. 2 is affected by the difficulty of knowing the electric field established in the tissue. On the other hand, temperature measurements (in eq. 14) correlated to dielectric properties (in eq. 2) allow an easy estimation of the electromagnetic field / tissue coupling intensity.

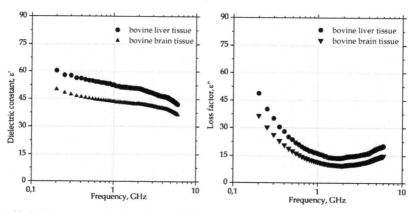

Figure 15. Dielectric constant and loss factor of liver and brain bovine tissues.

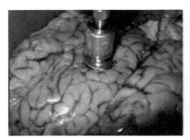

Figure 16. Snapshots of bovine tissues (brain and liver) dielectric properties measurements by open-end of the coaxial probe.

5. Modeling of microwave assisted processes

Modeling of microwave assisted processes can be classically approached starting from balance equations of momentum, energy and mass applied to a system under investigation and from solution of electromagnetic field distributions equations (Maxwell equations) (Acierno, Barba & d'Amore, 2008). With respect to the selected system a control volume, V, and a close surface, S, must be defined. Thus, the balance equations above cited, expressed through the general form of eq. 11 (Bird, Stewart and Lighfoot, 2002):

$$
\begin{Bmatrix} \text{rate of ACCumulation} \\ \text{of momentum, energy,} \\ \text{matter in the volume V} \end{Bmatrix} = \begin{Bmatrix} \text{INput flow rate} \\ \text{of momentum, energy,} \\ \text{matter through} \\ \text{the surface } S \end{Bmatrix} -
$$
$$
- \begin{Bmatrix} \text{OUTput flow rate} \\ \text{of momentu, energy,} \\ \text{matter through} \\ \text{the surface } S \end{Bmatrix} + \begin{Bmatrix} \text{GENeration rate} \\ \text{of momentum, energy,} \\ \text{matter} \\ \text{in the volume } V \end{Bmatrix} \tag{15}
$$

can be written. Applying the eq. 15 to the momentum $(\rho \underline{v})$, the equation of motion[5] is achieved:

$$
\frac{\partial}{\partial t}\rho \underline{\mathbf{v}} = -\underline{\nabla} \cdot \underline{\underline{\phi}} + \rho \mathbf{g} = -\nabla p - \underline{\nabla} \cdot \underline{\underline{\tau}} - \underline{\nabla} \cdot \rho \underline{\mathbf{v}}\underline{\mathbf{v}} + \rho \underline{\mathbf{g}} \tag{16}
$$

where t is the time variable [s], $\underline{\underline{\phi}}$ is the total flux of the momentum [N/m³], ρ is the density [kg/m³] and $\underline{g}$ is the gravity acceleration [m/s²]. The terms at the last member are the contributes of pressure (∇p), viscosity $(\underline{\nabla}\cdot\underline{\underline{\tau}})$ and convective $(\underline{\nabla}\rho\underline{v}\underline{v})$ fluxes. The generation term in the momentum balance equation is due to the gravity per unit volume. Reorganizing eq. 16 introducing the substantial derivative and under some conditions

---

[5] Notation: a scalar; $\underline{a}$ vector; $\underline{\underline{a}}$ tensor; $\cdot$ scalar product; x vectorial product; $\underline{\nabla}a$ gradient; $\underline{\nabla}\cdot\underline{a}$ divergence: ∇^2 a Laplacian.

(density as a function of temperature according to the Boussinesq approximation, with β volumetric expansivity coefficient; Newtonian flux, with μ dynamic viscosity [Pa·s]) the equation of motion becomes:

$$\rho_0 \frac{D}{Dt}\underline{v} = \left(-\nabla p + \rho_0 \underline{g}\right) - \mu \nabla^2 \underline{v} - \rho_0 \underline{g}\beta\left(T - T_0\right) \tag{17}$$

Eq. 17 is therefore the most useful form of momentum balance equation; it is solved introducing initial and boundary conditions and its solution represents the velocity field $\underline{v}(t, \underline{x})$, where $\underline{x}$ is the spatial vector.

Applying the eq. 15 to the energy $\rho\left(\hat{U} + \frac{1}{2}v^2\right)$, the equation of energy is achieved:

$$\frac{\partial}{\partial t}\rho\left(\hat{U} + \frac{1}{2}v^2\right) = -\nabla \cdot \underline{e} + \rho \underline{v} \cdot \underline{g} + GEN =$$
$$= -\nabla \cdot \rho\left(\hat{U} + \frac{1}{2}v^2\right)\underline{v} - \nabla \cdot \underline{q} - \nabla \cdot \left(\underline{\pi} \cdot \underline{v}\right) + \rho \underline{v} \cdot \underline{g} + GEN \tag{18}$$

Eq. 18 is the energy equation written using the concept of the total flow of energy($\nabla \cdot \underline{e}$), and making explicit the terms due to convection $\nabla \cdot \rho\left(\hat{U} + \frac{1}{2}v^2\right)\underline{v}$, conduction ($\nabla \cdot \underline{q}$) and viscous phenomena ($\nabla \cdot (\underline{\pi} \cdot \underline{v})$), where $\underline{\pi}$ accounts for pressure and viscous stress contributions to $\underline{\phi}$). The term ($\rho \underline{v} \cdot \underline{g}$) takes into account the generation of energy due to the motion against gravity (potential energy); other energy generation phenomena are included in the term GEN and they can refer to chemical / nuclear reactions, heat dissipation due to external causes (electrical, magnetic). Reorganizing eq. 18 introducing the substantial derivative and under some conditions (viscous stress is relevant only for high shear rates; processes at constant pressure, material density constant or slightly variable; Fourier conductive flux, with K thermal conductivity [W/m·K]), the equation of energy can be express by:

$$\rho \hat{C}_p \frac{D}{Dt}T = -K\nabla^2 T + \sum_i \Delta H_i r_i + \sum_j \dot{Q}_j \tag{19}$$

where $\Sigma_i \, \Delta H_i r_i$ and $\Sigma_j Q_j$ are the summations of generation terms due to phase changes/ chemical reactions (ΔH_i, latent heat, r_i volumetric rate) and to supplied/subtracted heat due to external sources (Q_j volumetric heat rate). The solution of eq. 19, performed after initial and boundary conditions definition, gives the temperature field $T(t, \underline{x})$.

Applying the eq. 15 to the mass ($\varrho_A = \rho \omega_A$) the mass equation is achieved:

$$\frac{\partial}{\partial t}\rho \omega_A = -\nabla \cdot \underline{n}_A + r_A = -\nabla \cdot \rho \omega_A \underline{v} - \nabla \cdot \underline{j}_A + r_A \tag{20}$$

where ω_A is the A component mass fraction. Eq. 20 is the mass equation, written using the concept of the total matter flow ($\nabla \cdot \underline{n}_A$) and making explicit the terms due to convection ($\nabla \cdot \rho \omega_A \underline{v}$) and diffusion ($\nabla \cdot \underline{j}_A$) phenomena. The generation term r_A expresses the volumetric rate of the A component appearance/disappearance. The solution of eq. 20, performed after initial an boundary conditions definition, gives the fraction mass field, $\omega_A(t, \underline{x})$. Under some condition (binary system with constant density; Fick diffusive flux, with D_{AB} diffusivity [m²/s]), the eq. 20 can be express by:

$$\frac{\partial}{\partial t}\rho_A + \underline{\mathbf{v}} \cdot \nabla \rho_A = D_{AB} \nabla^2 \rho_A + r_A \tag{21}$$

Maxwell equations in the frequency domain can be written by (Metaxas and Meredith, 1983; Chan & Reader, 2000; Meredith, 2006):

$$\begin{cases} \nabla \times \underline{E}(\underline{x},\omega) = -j\omega \underline{B}(\underline{x},\omega) - \underline{J}_m(\underline{x},\omega) - \underline{J}_{m0}(\underline{x},\omega) \\ \nabla \times \underline{H}(\underline{x},\omega) = j\omega \underline{D}(\underline{x},\omega) + \underline{J}(\underline{x},\omega) + \underline{J}_0(\underline{x},\omega) \\ \nabla \cdot \underline{D}(\underline{x},\omega) = \rho(\underline{x},\omega) + \rho_0(\underline{x},\omega) \\ \nabla \cdot \underline{B}(x,\omega) = \rho_m(\underline{x},\omega) + \rho_{m0}(\underline{x},\omega) \end{cases} \tag{22}$$

where $\underline{x}$ is, as usual, the position vector, ω is angular frequency (it takes in account the time variable) and where $\underline{E}(\underline{x}, \omega)$, $\underline{H}(\underline{x}, \omega)$ $\underline{D}(\underline{x}, \omega)$ $\underline{B}(\underline{x}, \omega)$ are le solutions for the electric field [V/m], magnetic field [A/m], induction electric field [C/m²] induction magnetic fields [Wb/m²] respectively. The other terms in eq. 22 are referred to the sources of the electromagnetic phenomena: $\underline{J}(\underline{x}, \omega)$ current electric density [A/m²]; $\underline{J}_m(\underline{x}, \omega)$ magnetic current density [V/m²]; $c(\underline{x}, \omega)$ electric charge density [C/m³]; $\rho_m(\underline{x}, \omega)$ magnetic charge density [Wb/m³]; $\underline{J}_0(\underline{x}, \omega)$ current electric density (external sources) [A/m²]; $\underline{J}_{m0}(\underline{x}, \omega)$ magnetic current density (external sources) [V/m²]; $\rho_0(\underline{x}, \omega)$ electric charge density (external sources) [C/m³]; $\rho_{m0}(\underline{x}, \omega)$ magnetic charge density (external sources) [Wb/m³].

Thus, the generic problem of electromagnetism, described by eq. 22 consists in solving a system of differential equations with four unknowns vector functions (the fields above defined). In general, to reduce the complexity of the problem the fields are linked to the inductions by relations (which are generally non-differential) dependent on the material in which the phenomena occur. These equations are called constitutive relations. Of great practical importance is the case of the media isotropic and temporally non-dispersive, for which it can be written:

$$\begin{cases} \underline{D}(\underline{x},\omega) = \varepsilon(\underline{x},\omega)\underline{E}(\underline{x},\omega) \\ \underline{B}(\underline{x},\omega) = \mu(\underline{x},\omega)\underline{H}(\underline{x},\omega) \end{cases} \tag{23}$$

where the functions $\varepsilon(\underline{x}, \omega)$ and $\mu(\underline{x}, \omega)$ are the complex permittivity and permeability, respectively. Their values can be achieved from direct measurements using the techniques discussed in previous paragraphs.

Summarizing, description of a process in which the material interacts with microwaves, requires the solution of the equations of change of momentum, energy and matter, in addition to the solution of the equations of the electromagnetic field (Maxwell's equations).

These two classes of equations are related through the material constants (viscosity, density, specific heat, thermal diffusivity, electric permittivity and magnetic permeability). In addition, crucial point in the problems of dielectric heating, is the evaluation of the heat generation term that appears in the energy balance equation due to dissipation of the microwave energy in the material. It is well evident that this is a very challenging task, dealing with partial differential equations whose solutions may show a functional dependency from four independent variables (time and three spatial coordinates). To make less burdensome the calculation procedures, problems can be suitably simplified prior to their resolution. The three types of approximations more broadly applied are (Bird, Stewart and Lighfoot, 2002):

i.   postulates, in which a tentative guess is made as to the form of the solution (as an example: the temperature field $T(t, x, y, z)$ can be expressed by $T(t, x)$);
ii. assumptions, in which one can eliminate some physical phenomena or effects by discarding terms or assuming physical properties to be constant;
iii. search for an asymptotic solution, in which one can obtain only a portion of the entire mathematical solution.

In balance equations, the phenomena that are not dominant can be neglected and/or material features can be kept as constant value. In the description of electromagnetic waves propagation in material media, simplified ways to describe the resolution of Maxwell's equations are constituted by the law of Lambert and Beer and the theories of waveguides and reverberant cavities. In the first case, it is possible to image electromagnetic waves as plane waves, whose intensity exponentially decreases along the direction of penetration (z) of the body with the following mathematical structure:

$$E(z) = E_0 \exp\left(-\frac{z}{D_p}\right) = E_0 \exp\left(-\frac{2\pi f}{c} \frac{\varepsilon''}{\sqrt{\varepsilon'}} z\right) \tag{24}$$

where z is the coordinate in the direction of propagation, E_0 is the intensity of the incident field; and where the D_p parameter depends from material dielectric properties (eq. 4).

In the second case it is possible to apply some simplifications of the Maxwell equations, consistent with the propagation of electromagnetic waves in isotropic systems, without electromagnetic field sources and structured as "pipes" (waveguides) or "closed cavity" (the reverberant cavity) made of conductive materials. Each of these solutions is a "mode" of propagation of the microwaves in the guides or in the cavities. The analytical forms of descriptive ways have been obtained and are summarized in electromagnetic handbooks. The modes of propagation are characterized by having Transverse Electric field (TE modes) or Transverse Magnetic field (TM modes), where the adjective transverse indicates field orthogonality to the axis of the guide or to the major axis of the cavity (Metaxas and Meredith, 1983; Chan & Reader, 2000).

If attention is focused on the mass and heat transfer phenomena occurring through wet materials (foods, pharmaceuticals) undergoing to a microwave assisted drying process, a simplified modeling of the overall process can be developed as follows. Starting from eq. 17, 19, and 21 and under the approximations:

1. no motion in the solid phase occurs, $\underline{v} = \underline{0}$;
2. the functions of interest (residual moisture, M, and temperature, T, profiles) only depend on time (t) and depth (z), i.e. $T = T$ (t, z) and $M = M$ (t, z);
3. liquid water does not diffuse in the solid matrix, $D_{AB} = 0$;
4. the following physical characteristics of the materials are considered constant: density (ρ), specific heat (Cp) and thermal conductivity (K) of the solid, latent heat of evaporation (ΔH_M) of the water;
5. microwaves (of a given frequency) propagate as a plane wave partially absorbed by the material according to the law of Lambert and Beer;

the variation equations of temperature and mass become:

$$\rho C_p \frac{\partial}{\partial t}T = -K\frac{\partial^2}{\partial z^2}T + \Delta H_M r_M + \dot{Q}_g \tag{25}$$

$$\frac{\partial}{\partial t}M = -r_M \tag{26}$$

The two mono-dimensional transient equations are coupled by the volumetric rate of water disappearance r_M. To solve the two partial differential equations, the following initial and boundary conditions can be chosen:

Initial Conditions:	@$t = 0$,"z	$T(0, z) = T_0, M(0, z) = M_0$ (27)
Boundary Conditions:	@$z = 0$,"t	$-K\frac{\partial T}{\partial z} = h\left(T\big|_{z=0} - T_\infty\right)$ (28)
	@$z = L$,"t	$\frac{\partial T}{\partial z} = 0$ (29)

where h ($T_{z=0} - T_\infty$) is the convective flux at the material surface, L is the thickness of the irradiated system. The energy dissipation due to microwaves is taken into account in the term $\dot{Q}_g$ of eq. 25, where the electric field strength is described using eqq. 2, 4 and 24 (summarized in eq. 30) with the assumption that microwaves propagate as a plane wave:

$$\dot{Q}_g = \frac{1}{2}\omega\varepsilon_0\varepsilon''(T,M)\left|E_0\exp\left(-\frac{2\pi f}{c}\frac{\varepsilon''(T,M)}{\sqrt{\varepsilon'(T,M)}}z\right)\right|^2 \tag{30}$$

Finally, dielectric properties correlation $\varepsilon(T, M) = \varepsilon(T(t,z), M(t,z))$ values must be known. At last, as one can be seen, experimental measurements of dielectric properties do play a key role. Other applications are reported in Barba et al., 2004, Acierno et al., 2008, Barba et al., 2012, Malafronte et al., 2012.

6. Conclusion

The main remarks of this chapter can be summarized as follows.

i. Microwave heating can be a powerful tool for thermal treatments because many benefits can be achieved but its successful use is directly associated with the dielectric properties of irradiated materials.

ii. In microwave assisted processes, the knowledge of dielectric properties and of parameters that affect their values allows to predict and provide desired heating partners in materials avoiding overheating and misheating.

iii. To measure dielectric properties, different techniques are available; their applicability are affected by several parameters such as intrinsic nature of materials under investigation (high or low dissipative features, solid or liquid state, magnetic or corrosive material, etc...); geometrical shape (thin film, thick solids, etc...); environmental conditions (high temperature, etc...). The open-ended coaxial-probe method is suitable to characterize many materials such as foods, pharmaceuticals and biological tissues.

iv. Modelling of microwave assisted processes can be approached by the solution of balance of heat, mass and momentum equations together with the solution of the electromagnetic field distributions equations. Complexity of the mathematical solutions of the coupled equations in their general form may be overcome taking advantage of the approximations possible when adapting the general problem to a peculiar application, with known geometry or properties, or predominant terms.

Author details

Anna Angela Barba and Matteo d'Amore
University of Salerno, Italy

Acknowledgement

This work was supported by the Ministero dell' Istruzione dell' Università e della Ricerca (contract grant number PRIN 2009 - 2009WXXLY2).

7. References

Acierno, D., Barba, A.A., & d'Amore M. (2008). I Fenomeni di Trasporto, In: *Il riscaldamento a microoonde. Principi ed Applicazioni*, pp. (53-75), Pitagora Editrice, ISBN 9-788837-116996, Bologna, IT

Adair, E.R., & Petersen, R.C. (2002). Biological Effects of Radio-Frequency/Microwave Radiation. *IEEE Transactions on Microwave Theory and Techniques*, Vol. 50, No. 3, (March 2002), pp. (953-962), ISSN 0018-9480

Al-Harahsheh, M.,& Kingman, S.W. (2004). Microwave-assisted leaching—a review. *Hydrometallurgy*, Vol. 7, No. 3-4, pp. (189–203), ISSN 0304-386X

Auriemma, G., Del Gaudio, P., Barba, A. A., d'Amore, M., & Aquino, R.P. (2011). A combined technique based on prilling and microwave assisted treatments for the production of ketoprofen controlled release dosage forms. *International Journal of Pharmaceutics* Vol. 415, No. 1-2, pp. (196–205), ISSN 0378-5173

Barba, A.A., Acierno D., & d'Amore M. (2012). Use of microwaves for in-situ removal of pollutant compounds from solid matrices. *Journal of Hazardous Materials*, Vol. 207-208, No. March, pp.(128-135), ISSN 0304-3894

Barba, A.A., & d'Amore M. (2008). Applicazioni delle microonde nel settore agroalimentare, In: *Il riscaldamento a microonde. Principi ed Applicazioni*, pp. (269-295), Pitagora Editrice, ISBN 9-788837-116996, Bologna, IT

Barba, A.A., Guidotti, C., & d'Amore M. (2010). Processi intensificati in ambito farmaceutico: le applicazioni di potenza delle microonde, *Notiziario Chimico-Farmaceutico*, Febb. 2010, pp. (68-71), ISSN 0393-3733- 6.

Barba, A.A., Notari A., & d'Amore, M. (2010). Tissue-equivalent materials in studying interactions between non-ionizing radiation and living systems, *Proceedings of XVIII RiNEm*, ISBN 978-88-905261-0-7 Benevento, Italy, September 6th–10th 2010

Barba, A.A., d'Amore, M., & Acierno, D. (2004). Heat Transfer Phenomena During Processing Materials with Microwave Energy. *Heat and Mass Transfer*, Vol. 40; No. 5, pp. (413-420), ISSN 0947-7411

Bird, R.B., Stewart, W.E., & Lightfoot, E.N. (2002). *Transport Phenomena* (2nd Ed.) John Wiley & Sons, ISBN 0-471-36474-6, Danvers, MA-USA

Chan, C.T.T.V. & Reader, H.C. (2000). *Understanding Microwave Heating Cavities*, Artech House Inc., ISBN 158053094-X Norwood, MA-USA

Coulson & Richardson's, Chemical Engineering (2002). Product Design and Process Intensification, In: *Particle Technology & Separation Processes*, Richardson, J.F.; Harker, J.H.; Backhurst, J.R, 5th Ed., Vol. 2, pp. (1104-1135) Elsevier Butterworth Heinemann, ISBN 978-0-7506-4445-7, NL

Foster, K.R., & Adair, E.R. (2004). Modeling thermal responses in human subjects following extended exposure to radiofrequency Energy. *BioMedical Engineering OnLine*, Vol. 3:4, No. February, pp. (1-7)

Gabriel, C., Gabriel, S., & Corthout, E. (1996). The dielectric properties of biological tissues: I. Literature survey. *Physics in Medicine and Biology*, Vol. 41, No. 11, pp. (2231–2249) ISSN 0031-9155

Gabriel, S., Lau, R.W., & Gabriel, C. (1996). The dielectric properties of biological tissues: II. Measurements in the frequency range 10 Hz to 20 GHz. *Physics in Medicine and Biology*, Vol. 41, No. 11 , pp. (2251–2269) ISSN 0031-9155

Gabriel, S., Lau, R.W., & Gabriel, C. (1996). The dielectric properties of biological tissues: III. Parametric models for the dielectric spectrum of tissues. *Physics in Medicine and Biology*, Vol. 41, No. 11, pp. (2271–2293) ISSN 0031-9155

Gradinarsky, L., Brage, H., Lagerholm B., Björn, I.N., & Folestad, S. (2006). In situ monitoring and control of moisture content in pharmaceutical powder processes using an open-ended coaxial probe. *Measurement Science and Technology*, Vol. 17, pp. (1847-1853) ISSN 0957-0233

Guy, A.W. (1971). Analyses of electromagnetic fields induced in biological tissues by Thermographic Studies on Equivalent Phantom Models. *IEEE Transaction on Microwave Theory and Techniques*, Vol. mtt-19, No. 2, pp. (205 - 214), ISSN: 0018-9480

Hegedus, A., & Pintye-Hodi, K. (2007). Comparison of the effects of different drying techniques on properties of granules and tablets made on a production scale. *International Journal of Pharmaceutics*, Vol. 330, No. 1-2, pp. (99–104) ISSN: 0378-5173

Heng, P.W.S., Loh, Z.H., Liew, C.V., & Lee, C.C. (2010). Dielectric properties of pharmaceutical materials relevant to microwave processing: effects of field frequency, material density, and moisture content. *Journal of Pharmaceutical Sciences*, Vol. 99, No. 2, pp. (941–957), ISSN 1520-6017

Kelen, A., Ress, S., Pallai E., & Pintye-Hòdi, K. (2006). Mapping of temperature distribution in pharmaceutical microwave vacuum drying. *Powder Technology*, Vol. 162, pp. (133-137) ISSN 0032-5910

Kraszewski, A. (1996). *Microwave Aquametry Electromagnetic Wave Interaction with Water-Containing Materials*, IEEE Press, ISBN 0-7803-1146-9, New York, USA .

Kulkarni, P.S., Crespo, J.G., & Afonso, C.A.M. (2008). Dioxins sources and current remediation technologies — a review. *Environment International*, Vol. 34, No. 1, pp. (139–153), ISSN 0160-4120

Loh, Z.H., Liew, C.V., Lee, C.C., & Heng, P.W.S. (2008). Microwave-assisted drying of pharmaceutical granules and its impact on drug stability. *International Journal of Pharmaceutics*,Vol. 359, No. 1-2, pp. (53–62), ISSN: 0378-5173

Malafronte, L., Lamberti G., Barba A.A., Raaholt B., Holtz E., & Ahrné, L. (2012), Combined convective and microwave assisted drying: experiments and modeling. *Journal of Food Engineering*, Vol. 112, No. 4, pp. (304–312) ISSN 0260-8774

McKeown, M.S., Trabelsi S., Tollner, E.W., & Nelson S.O. (2012). Dielectric spectroscopy measurements for moisture prediction in Vidalia onions. *Journal of Food Engineering*, Vol. 111, No. 3, pp. (505–510), ISSN 0260-8774

McMinn, W.A.M., McLoughlin T.C.M., & Magee, T.R.A. (2005). Microwave–convective drying characteristics of pharmaceutical powders. *Powder Technology*, Vol. 153, No. 1, pp. (23– 33), ISSN 0032-5910

Meredith, R.J. (1998). *Engineers' Handbook of Industrial Microwave Heating*, IEE Power Engineering, ISBN-10: 0852969163, London, UK

Metaxas, A.C. & Meredith, R.J. (1983). *Industrial Microwave Heating*, Peter Peregrinus Ltd., ISBN 0906048893, London, UK

Mudgett, R.E. (1986). Electrical properties of foods, In: *Engineering properties of foods*, Rao M.A. and Rizvi S.S.H., Eds., Marcell Dekker Inc., ISBN 0-8247-7526-0, pp. (49-87), New York, 1986

Nurjaya, S, & Wong T.W.(2005). Effects of microwave on drug release properties of matrices of pectin. *Carbohydrate Polymer*, Vol. 62, No. , pp. (245-257) ISSN 0144-8617

Olhsson, T. (2000). Minimal processing of foods with thermal methods, In: *Innovations in Food Processing*, Barbosa-Canovas G.V., Gould G.W., Eds., pp. (141-148) Technomic Inc., ISBN 1-56676-782-2, Lancaster, PA, USA

Remya, N., & Lin, J.G. (2011). Current status of microwave application in wastewater treatment—a review. *Chemical Engineering Journal*, Vol. 166, No. 3, pp. (797–813), ISSN 1385-8947

Schubert, H., & Regier M. (2005). *The Microwave Processing of Foods*, CRC Press Woodhead Publishing, ISBN 978-1-85573-964-2 Boca Raton, FL, USA

Sipahioglu, O., & Barringer, S.A. (2003). Dielectric Properties of Vegetables and Fruits as a Function of Temperature, Ash, and Moisture Content. *Journal of Food Science*, Vol. 68, No. 1, pp. (234–239), ISSN 1750-3841

Stankiewicz, A. & Moulijn J. (2004). *Re-engineering the chemical processing plant: process intensification*, M. Dekker, ISBN 0824743024 9780824743024 New York, USA

Tang, J., Hao, F., & Lau, M. (2002). Microwave Heating in Food Processing, In: *Advanced Bioprocessing Engineering*, X.H. Yang and J. Tang Eds., pp. (1-44), Word Scientific Ltd., ISBN 981-02-4696-X, Singapore

Wang, S., Tang, J., Johnson, J.A., Mitcham, E., Hansen, J.D., Hallman, G., Drake, S.R., & Wang, Y. (2003). Dielectric Properties of Fruits and Insect Pests as related to Radio Frequency and Microwave Treatments. *Biosystems Engineering*, Vol. 85, No. 2, pp. (201–212), ISSN 1537-5110

Wong, T.W., Chan L.W., & Kho, S.B., & Heng, P.W.S. (2002). Design of controlled release solid dosage forms of alginate and chitosan using microwave. *Journal of Control Release*, Vol. 84, No. 3, pp. (99-114), ISSN 0168-3659

Zhou, J., Shi, C., Mei, B., Yuan, R., & Fu, Z. (2003). Research on the technology and the mechanical properties of the microwave processing of polymer. *Journal of Materials Processing Technology*, Vol. 137, No. 1-3, pp. (156–158), ISSN 0924-0136

Free-Space Transmission Method for the Characterization of Dielectric and Magnetic Materials at Microwave Frequencies

Irena Zivkovic and Axel Murk

Additional information is available at the end of the chapter

1. Introduction

Materials that absorb microwave radiation are in use for different purposes: in anechoic chambers, for electromagnetic shielding, in antenna design, for calibration targets of radiometers, etc. It is very important to characterize them in terms of frequency dependent complex permittivity and permeability for a broad frequency range.

Widely used absorbing materials are CR Eccosorb absorbers from Emerson&Cuming Company. Permittivity and permeability of these materials are characterized by manufacturer up to frequency of 18GHz, but it is important (in absorbing layer design purposes, for example) to know these values at much higher frequencies.

In this work we will present new method, retrieved results and validation for complex and frequency dependent permittivity and permeability parameter extraction of two composite, homogeneous and isotropic magnetically loaded microwave absorbers (CR Eccosorb). Permittivity and permeability are extracted from free space transmission measurements for frequencies up to 140GHz. For the results validation, reflection measurements (samples with and without metal backing) are performed and are compared with simulations that use extracted models. The same method is applied in complex and frequency dependent permittivity model extraction of commercially available epoxies Stycast W19 and Stycast 2850 FT.

The proposed new method solves some shortcomings of the popular methods: extracts both permittivity and permeability only from transmission parameter measurements, gives good results even with noisy data, does not need initial guesses of unknown model parameters.

2. Definition of permittivity and permeability

In the absence of dielectric or magnetic material, there are the following relations:

$$D = \varepsilon_0 \cdot E \tag{1}$$

$$B = \mu_0 \cdot H \tag{2}$$

where D is electric induction, E is electric field, B is magnetic induction and H is magnetic field. ε_0 and μ_0 are permittivity and permeability of free space, respectively. Values of ε_0 and μ_0 are: $\varepsilon_0=8.854 \cdot 10^{-12}$ [$\frac{F}{m}$] and $\mu_0=4 \cdot \pi \cdot 10^{-7}$ [$\frac{V \cdot s}{A \cdot m}$]. Dielectric permittivity and magnetic permeability of the free space are related to each other in the following way:

$$c^2 = \frac{1}{\mu_0 \cdot \varepsilon_0} \tag{3}$$

where c is the speed of light in a vacuum and its value is $c \approx 3 \cdot 10^8$ [$\frac{m}{s}$].

If an electromagnetic field interacts with material that is dielectric or magnetic, equations (1) and (2) can be represented as follows:

$$D = \varepsilon_0 \cdot \varepsilon \cdot E \tag{4}$$

$$B = \mu_0 \cdot \mu \cdot H \tag{5}$$

where ε and μ are relative permittivity and permeability of the observed material and can be real or complex numbers (Eq. (6) and (7)).

$$\varepsilon = \varepsilon' - j \cdot \varepsilon'' \tag{6}$$

$$\mu = \mu' - j \cdot \mu'' \tag{7}$$

where j is imaginary unit and j^2= -1.

Permittivity is a quantity that is connected to the material's ability to transmit ('permit') an electric field. The real part, ε', is related to the ability of material to store energy, while the imaginary part, ε'', describes losses in material. Permeability is a parameter that shows the degree of magnetization that a material obtains in response to an applied magnetic field. Analogous to the real and imaginary permittivity, the real permeability, μ', represents the energy storage and the imaginary part, μ'', represents the energy loss term.

The polarization response of the matter to an electromagnetic excitation cannot precede the cause, so Kramers-Kronig relations given by Eqs. (8) and (9) ([8]), for the real and imaginary parts of permittivity (permeability) have to be fulfilled.

$$\varepsilon'(\omega) = \varepsilon_\infty + \frac{2P}{\pi} \int_0^\infty \frac{\omega' \varepsilon''(\omega')}{\omega'^2 - \omega^2} d\omega' \tag{8}$$

$$\varepsilon''(\omega) = -\frac{2\omega P}{\pi} \int_0^\infty \frac{\varepsilon'(\omega') - \varepsilon_\infty}{\omega'^2 - \omega^2} d\omega' \tag{9}$$

where ω is angular frequency, ε_∞ is permittivity when $\omega \to \infty$ and P stands for the Cauchy principal value. The convention of the time variation is $\exp(j\omega t)$ and the time derivative is equal to multiplication by $j\omega$.

The Kramers-Kronig relations connect the real and imaginary parts of response functions. If we know the real/imaginary part of permittivity/permeability on the complete frequency range, the other unknown part can be calculated using the Kramers-Kronig relations. Because of the causality constraint, when we develop models for frequency dependent permittivity or permeability, they must satisfy the Kramers-Kronig relations.

3. Modeling of frequency dependent permittivity and permeability

Dielectric and magnetic loss mechanisms can be represented in the frequency domain as relaxation or resonant type. In the microwave frequency range dielectric losses usually exhibit relaxation behavior ([1]), while magnetic losses exhibit resonant behavior ([6]), ([7]).

3.1. Debye relaxation model

The simplest one pole Debye relaxation model is represented with the Eq. (10).

$$\varepsilon(f) = \varepsilon_\infty + \frac{\varepsilon_s - \varepsilon_\infty}{1 + j\frac{f}{f_r}} \tag{10}$$

where ε_s is a static dielectric permittivity, ε_∞ is permittivity at infinite frequency (optical permittivity), f_r is relaxation frequency. Figure 1 gives an example of Eq. (10) with $\varepsilon_\infty = 7$, $\varepsilon_s = 17$ and $f_r = 9$ (in GHz). Imaginary part of permittivity in Figure 1 is represented as a positive number.

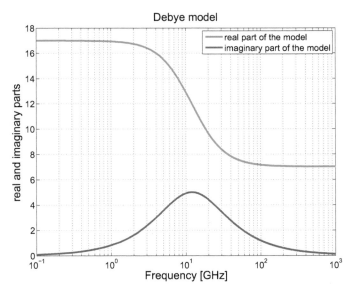

Figure 1. Real and imaginary permittivity represented with Debye model.

There are some modified Debye relaxation models that include asymmetrical and damping factors ([2]). These models are Cole-Cole, Cole-Davidson and Havriliak-Negami. They are presented with equations (11), (12) and (13), respectively.

$$\varepsilon(f) = \varepsilon_\infty + \frac{\varepsilon_s - \varepsilon_\infty}{1 + j\left(\frac{f}{f_r}\right)^{1-\alpha}} \tag{11}$$

$$\varepsilon(f) = \varepsilon_\infty + \frac{\varepsilon_s - \varepsilon_\infty}{\left(1 + j\left(\frac{f}{f_r}\right)\right)^{\beta}} \tag{12}$$

$$\varepsilon(f) = \varepsilon_\infty + \frac{\varepsilon_s - \varepsilon_\infty}{\left(1 + j\left(\frac{f}{f_r}\right)^{1-\alpha}\right)^{\beta}} \tag{13}$$

The terms α and β are empirical parameters and their values are between 0 and 1. α is a damping factor and describes the degree of flatness of the relaxation region. β is an asymmetric factor and describes relaxation properties asymmetric around relaxation frequency.

In our work, we will model dielectric permittivities of the samples with Debye relaxation model given with Eq. (10).

3.2. Lorenzian resonance model

Lorenzian resonant model is represented with Eq. (14) and Eq. (15). Graphical representation of Eq. (14) is in Figure 2 which is an example with $\mu_s = 9$ and $f_r = 25$ GHz. Imaginary part of permeabilty in Figure 2 is represented as positive number.

$$\mu(f) = 1 + \frac{\mu_s - 1}{\left(1 + j\frac{f}{f_r}\right)^2} \tag{14}$$

where μ_s is static permeability and f_r is resonant frequency.

$$\mu(f) = 1 + \frac{\mu_s - 1}{1 + j\frac{f}{f_{r1}} - \left(\frac{f}{f_{r2}}\right)^2} \tag{15}$$

Eq. (15) comes when we develop Eq. (14). If $2f_{r1} = f_{r2}$, Eq. (14) is equivalent to the Eq. (15).

Damping and asymmetric factors are introduced in the following equations ([2]).

$$\mu(f) = 1 + \frac{\mu_s - 1}{1 + j\gamma\frac{f}{f_r} - \left(\frac{f}{f_r}\right)^2} \tag{16}$$

$$\mu(f) = 1 + \frac{\mu_s - 1}{\left(1 + j\gamma\frac{f}{f_r} - \left(\frac{f}{f_r}\right)^2\right)^k} \tag{17}$$

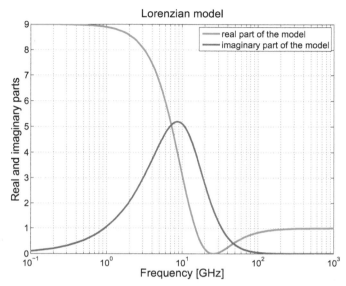

Figure 2. Real and imaginary permeability represented with Lorenzian model.

The γ factor is an empirical constant and represents damping factor of a resonance type. The term k is also an emirical constant which values are between 0 and 1. It is asymmetrical factor of a resonance type.

In our work we will model complex permeability with Lorenzian resonant model and we will 'tune' models by involving empirical factors k and γ.

4. Scattering parameters measurements and methods for permittivity and permeability extractions

When an electromagnetic wave interacts with a material sample of finite thickness, reflected and transmitted signals can be registered. On the air-material sample interface, part of the signal reflects back, while the other part penetrates into the material. Inside the material the signal attenuates and on the second material sample-air interface part of the signal reflects back into the sample while the other part continues to go forward, ie. transmits. Scattering parameters (transmission and reflection) depend on permittivity (ε) and permeability (μ) of the material and by measuring them we can extract ε and μ.

4.1. Transmission/reflection waveguide measurements

Coaxial line, rectangular and cylindrical waveguide measurements are widely used broadband measurement techniques because of their simplicity ([5]), ([3]). In these methods, the material sample is placed into the section of a waveguide or a coaxial line and scattering parameters are measured with network analyzer.

A major problem in transmission line measurements is the possible existance of air gaps between the sample and walls of the waveguide. Samples must be very precisely machined. For example, in X band waveguide measurements (8 to 13GHz, waveguide dimensions 22.9 x 10.2mm), uncertainty in the sample cutting should be in the range of $20\mu m$ ([5]).

Important problems with transmission line measurements are also half wavelength resonance and overmoding ([5]). Propagation of single mode resonates at integer multiplication of one half wavelength in the sample. Some techniques for permittivity and permeability extractions break down in the presence of half wavelength resonance. An additional problem is overmoding which means appearance and propagation of higher order modes in the closed waveguide structure. In waveguides and coaxial lines the asymmetry of the sample and machining precision promotes higher order mode propagation. Their appearance determines dips in the reflection coefficient caused by resonances of the excited higher order mode which cause failure of, for example, point by point technique for permittivity and permeability extraction.

4.1.1. Free space measurements

Free space measurements give a noninvasive broadband technique for transmission and reflection parameters measurements. Scattering parameters are measured of the sample that is plane parallel. Measurement setup consists of two identical antennas that operate in certain frequency range and network analyzer. For measurements, corrugated horn antennas can be used. Antennas are aligned and one of them transmits signal while the other antenna works as receiver. Material sample is placed between the two antennas, the incident signal passes through material and is registered by the other antenna. On that way, free space transmission coefficient is measured. Depending on frequency and sample size, focusing lenses can be used.

For reflection measurements, one antenna is connected to the network analyzer via directional coupler. The antenna is sending signal and also measures reflection from the sample that is in front of the antenna.

4.2. Methods for permittivity and permeability extractions

4.2.1. Analytical approach - Nicholson Ross Weir derivation

The Nicholson Ross Weir (NRW) derivation is an analytical method that calculates permittivity and permeability from measured S_{11} and S_{21} parameters. Dependence of scattering parameters from material properties is derived considering multiple reflections of the wave incident upon the air-sample interfaces when the sample is in free space or inside of waveguide ([5]), ([3]). Equations (18) to (29) represent short version of the NRW derivation for scattering parameters measured in free space. If we consider a system of air/sample/air, then the incident wave travels and a first partial reflexion on the air-sample interface occurs. The remaining portion of the signal continues to travel through the sample and on the second air-sample interface part of the signal transmits and the other part reflects back and travel through the sample toward the first air-sample interface. After simplification of expressions that include all terms of multiple reflections and transmissions, the final expression for the total reflection parameter, S_{11}, and the total transmission parameter, S_{21}, are given with Eq. (18) and Eq. (19).

$$S_{11} = \frac{G1 \cdot (1 - z^2)}{(1 - G1^2 \cdot z^2)} \tag{18}$$

$$S_{21} = \frac{z \cdot (1 - G1^2)}{(1 - G1^2 \cdot z^2)} \tag{19}$$

$$z^2 = e^{-2 \cdot j \cdot \gamma \cdot d} \tag{20}$$

$$G1 = \frac{(Z - 1)}{(Z + 1)} \tag{21}$$

$$Z = \sqrt{\frac{\mu}{\varepsilon}} \tag{22}$$

where z is unknown variable that depends on the propagation constant γ, d is the sample thickness, $G1$ is the first partial reflection on the air-sample interface, Z is characteristic impedance of the material and depends on ε and μ of the material as given with Eq. (22). Relation of the propagation constant γ with ε and μ of the material is given by Eq. (23).

$$\gamma = \frac{j \cdot 2 \cdot \pi \cdot \sqrt{\varepsilon \cdot \mu}}{\lambda} \tag{23}$$

$$\lambda = \frac{c}{f} \tag{24}$$

where λ is a free space wavelength, c is speed of light in vacuum and f is the frequency.

$$\gamma = -\frac{1}{d} \cdot \log \frac{1}{z} \tag{25}$$

From the Eq. (20), γ is expressed as a function of z and d (Eq. (25)). d is material slab thickness and z is calculated from Eq. (18) and Eq. (19).

N_m is material's refractive index and can be expressed in terms of permittivity and permeability as:

$$N_m = \sqrt{\mu \cdot \varepsilon} \tag{26}$$

By combining Eq. (23) and Eq. (26) we obtain expression for material's refractive index (Eq. (27)). $G1$ is expressed as a funcion of S_{11} and S_{21} and Z is expressed as a function of $G1$. Finally, ε and μ are expressed as a functions of N_m and Z (Eqs. (28) and (29)).

$$N_m = \frac{-j \cdot \lambda \cdot \gamma}{2 \cdot \pi} \tag{27}$$

$$\varepsilon = \frac{N_m}{Z} \tag{28}$$

$$\mu = N_m \cdot Z \tag{29}$$

The numerical shortcoming of the NRW derivation comes from Eq. (25) where the natural logarithm of a complex number has to be calculated and there is no unique solution. The choice of the correct root is essential in order to find the correct solution of the complex permittivity and permeability. Comparison of the calculated and measured time delays can help in resolving this problem ([5]). The second problem arises when S_{11} and S_{21} are noisy. The NRW derivation calculates ε and μ from S_{11} and S_{21} for each frequency point. Small measurement errors in the dataset result in significant errors on the calculated values of ε and μ. Figure 4 represents an example of permittivity and permeability calculation with NRW algorithm when S_{11} and S_{21} are noisy. Frequency dependent permittivity and permeability represented with Debye and Lorenzian models are included into the Fresnel's equations based algorithm. Fresnel's equations based algorithm then calculates S_{11} and S_{21} (Figure 3). Noise with relative amplitude of 0.05 is added to simulated S_{11} and S_{21}. Figure 4 represents original ε and μ as well as ε and μ calculated with NRW derivations.

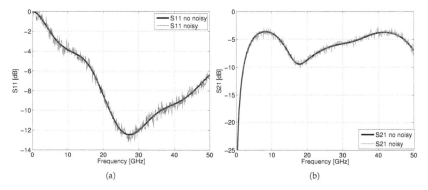

(a)   (b)

Figure 3. Noisy S_{11} and S_{21} that are used for permittivity and permeability calculations with NRW derivation.

4.2.2. Numerical approach

Taking various linear combinations of scattering parameters it is possible to calculate unknown permittivity and permeability ([5]). In order to obtain values for both parameters, we need to have two different measurements of scattering parameters: S_{11} and S_{21} measured on one material sample, or one port shorted circuit line measurement of material sample at two different positions in the line, or measurement of one scattering parameter but for the same material and two different thicknesses, etc. For the root determination of the equations (that are similar to the NRW equations) Newton's numerical method can be used ([5]). This iterative approach works well if good initial guesses are available.

One of the numerical methods for both permittivity and permeability determination is based on nonlinear least square optimization technique and is described in details in ([4]). Complex permittivities and permeabilities are represented as a sum of resonance and relaxation terms. Measured scattering parameters are fitted with simulations in the nonlinear least square sense and unknown free parameters from permittivity and permeability models are extracted. It is important to have a good initial guesses of the unknown parameters contained in the ε and μ models in order of optimization to converge to the correct solution. The initial guesses should

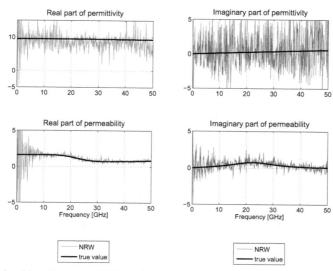

Figure 4. Real and imaginary permittivity and permeability calculated from noisy S_{11} and S_{21} by using
NRW derivation. Black line represents original values of permittivity and permeability which were used
to simulate the S-parameters.

be within 10 to 20% of the true values. The problem is when we examine material without any
a priori knowledge about its properties, then we are not able to give good initial guesses of its
unknown parameters.

5. Developed method for permittivity and permeability extraction

There are uncertainties in free space scattering parameters measurements and they have some
frequency dependence with higher frequencies having larger uncertainties. More sensitive to
measurement uncertainties is the S_{11} parameter ([5]), ([3]), ([4]). For that reason we use free
space transmission S_{21} parameter for permittivity and permeability model retrieval, and free
space reflection (S_{11} and S_{11m}, reflections from the samples with and without metal backing)
parameter for extracted permittivity and permeability models validation.

Measurements and parameters extraction are concerned Eccosorb CR110, CR114 and CRS117
samples. The CR materials are a two composite-mixture of epoxy and magnetic inclusions
and are not flexible materials. A higher product number indicates a higher filling factor of
the magnetic loading and therefore the higher absorption. The CRS materials are flexible
silicon based materials. They should have the same electrical properties as hard epoxy based
materials.

In our work we use the assumption that all material samples are two composite, homogeneous
and isotropic. Dimensions and densities of the samples that we examined are summarized in
the Table 1. The samples are 2.00mm thick with plane parallel circular surfaces.

Eccosorb material	Diameter $[mm]$	Density $[\frac{g}{cm^3}]$
CR110	91.70	1.60
CR114	92.80	2.88
CRS117	75.60	4.16

Table 1. CR samples dimensions and densities

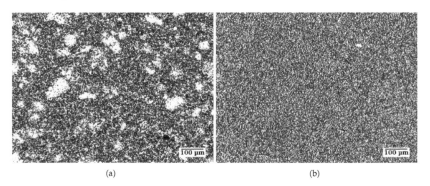

(a)   (b)

Figure 5. Microscope view of CR110 (left) and CR114 (right) samples. Photographs are taken with the same magnification for both samples.

Figure 5 represents microscopic view of CR110 and CR114 materials. Different densities of materials can be noticed. For these images, both samples had been polished to approximately 20μm thick slabs.

5.1. Scattering parameters measurements

5.1.1. Transmission measurements

Free space transmission measurements are performed with a setup explained in section 4.1.1. Measurements are performed at frequencies from 22 to 140GHz. Low frequency measurements are limited by the sample size.

Instead of placing material sample between two aligned antennas, we do transmission measurements by placing sample on the aperture of one of the antennas. Corrugated horn antennas are used because they have less near field effects at the aperture than rectangular horns. Calibration for S_{21} measurements is done with 'through' measurement Eq. (30). 'Through' signal is measured when antennas are separated only by air, without sample.

$$S_{21cal} = \frac{S_{21meas}}{S_{21through}} \qquad (30)$$

where S_{21cal} is the calibrated signal, S_{21meas} is the transmission parameter measured through material sample and $S_{21through}$ is the through measurement, with the signal received by the other antenna when no sample is between the two antennas.

5.1.2. Reflection measurements

Reflection measurements are performed by placing the material sample on the aperture of the antenna. We did two types of reflection measurements: with sample and metal backing (a metalic reflector placed behind the sample) placed on the aperture and only with the sample placed on the antenna's aperture.

The antenna is connected to the vector network analyzer through directional coupler. For calibration purposes we measure the reference signal (metal plate is on the aperture of the antenna, 100% reflection) and the signal when low reflectivity pyramidal foam absorber ($S_{11foam} < $ -50dB) is in front of the antenna's aperture (to calibrate directivity). Calibration of the measured reflection parameter is given by Eq. (31).

$$S_{11cal} = \frac{S_{11meas} - S_{11foam}}{S_{11alu} - S_{11foam}} \tag{31}$$

where S_{11cal} is the calibrated signal, S_{11meas} is the measured reflection parameter of the sample, S_{11foam} is the measured reflection when foam absorber is on the top of aperture and S_{11alu} is reference reflection measurement, when metal is on the aperture of the antenna.

5.2. Procedure for parameters extraction

We do not have any information about epoxy and magnetic inclusions properties. Our assumptions on the examined material samples are described as follows:

- the samples are two component composite materials (this is given by manufacturer, samples are mixtures of dielectric matrix and magnetic particles);
- inclusions are smaller when compared to the wavelength;
- the material is isotropic and homogeneous at macroscopic scale.

Epoxy is low loss dielectric material, while magnetic inclusions have magnetic and also dielectric properties. With previous assumptions and according to ([8]), if magnetic dispersion of inclusions is of resonance Lorenzian type than dispersion law for the composite will be Lorenzian as well. Analogous to that, dielectric dispersion of the composite can be modeled with relaxation Debye model.

At high frequencies (> 60-70GHz) there are no magnetic losses because magnetization is not possible since applied field is very fast and magnetic domains cannot follow the field. Permeability is equal to 1. It means that material samples exhibit only dielectric losses. We model dielectric losses with simple Debye model (Eq. (11)). Next step is to fit measured transmission parameter, with simulated S_{21}, both amplitude and phase. Fitting is based on minimization of the differences (in both amplitude and phase) between simulated and measured transmission data. Model of free space propagation is required to relate the material properties (permittivity and permeability) to the transmission (reflection) parameters. For that purpose we use routine based on Fresnel's equations. There are three unknowns in Debye model for permittivity calculation, ε_s, ε_∞ and f_r. Static permittivity, ε_s, is calculated for all samples by measuring capacitance of the sample at very low frequency. We measure the sample's capacitance in a calibrated capacity bridge operating between 10Hz and 20kHz. Capacitance measurement works good if wavelength is much longer then the sample thickness. It is satisfied in our case because capacitance measurements are performed at 1kHz frequency and samples thicknesses are 2mm. One problem in capacitance measurements is given by the fringing fields. To eliminate them, we measure capacitance of the sample C and then capacitance of the capacitor with air instead of material sample C_{air}. Static permittivity ε_s is expressed by Eq. (32).

$$\varepsilon_s = \frac{C}{C_{air}} \tag{32}$$

where C is the measured capacitance of the material sample and C_{air} is the capacitance between parallel capacitor plates which are separated for a distance equal to the thickness of the material sample, but instead of material there is air.

We include measured ε_s into permittivity Debye model. The next step is to do measured and simulated data fitting at high frequencies (μ=1), both amplitude and phase, and to extract two other unknown parameters of Debye model, f_r and ε_∞. Once we have full Debye dielectric permittivity model, we can extract permeability.

Starting guess is that permeability of the CR110 sample satisfies Debye relaxation model. As a matter of fact, CR110 is the sample with the smallest amount of magnetic inclusions so permeability behavior should change from resonance to relaxation. Permeability models of CR114 and CRS117 samples are presumed to be of Lorenzian type. With these guesses, we do fitting of measurements and simulations at frequencies where permeability is different than 1, while for permittivity we use a model extracted in the previous step (from fitting with high frequency data). From fitting, we obtain free parameters of presumed permeability models.

Figure 6 represents measured and fitted amplitude and phase of examined Eccosorb samples. A phase offset of one π (Figure 6 (b)) has been applied between phases of different samples for clarity.

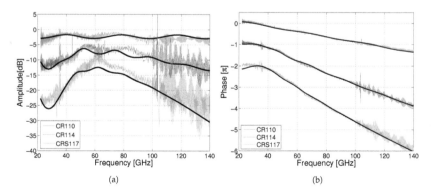

(a)   (b)

Figure 6. Measured and fitted transmission parameter amplitudes (a) and phases (b) of CR110, CR114 and CRS117 samples.

5.3. Extracted permittivity and permeability of CR Eccosorb absorbers and results validation

To validate extracted models for both permittivity and permeability, we compare simulated reflection parameters (samples with and without metal backing) with measurements. Comparisons are presented in Figures 7 and 8 and good agreement between measurements and simulations is achieved. Reflection measurements are performed in Ka, U and W band.

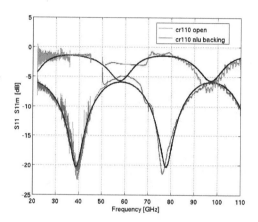

Figure 7. Measured and simulated reflection coefficient of CR110 absorber, with and without metal backing. Simulated reflections are represented with black solid line.

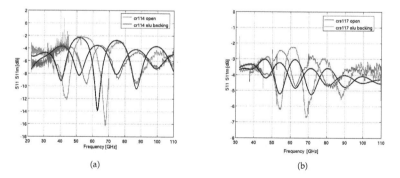

Figure 8. Measured and simulated reflection coefficient of CR114 (a) and CRS117 (b) absorbers, with and without metal backing. Simulated reflections are represented with black solid line.

There are some inconsistencies in measurements in different frequency bands. That could come from the fact that we used different corrugated horn antennas for different frequency bands. Another possible source for the inconsistency in measurements can be caused by the presence of the air gaps between samples and metal backing. Small air gaps exist because some of the examined samples are not completely flat, but are slightly bended.

By looking in amplitude behavior of transmission measurements (Figure 6 (a)) we can say that if permittivities and permeabilities are not frequency dependent, transmission coefficient decreases with increasing frequency. The fact that measured transmission coefficient decreases in some frequency range and increases in the other, says about frequency dependent material parameters.

Figures 9 and 10 represent extracted frequency dependent real and imaginary parts of permittivity and permeability of examined samples. As we mentioned, CR110 sample is low loss material with very small amount of magnetic particles that can produce losses. Because of the small concentration of magnetic particles, magnetic loss mechanism is transformed from resonance to relaxation (which is seen in Figure 10 (b), imaginary part of permeability). Both CR114 and CRS117 materials show Debye relaxation model for permittivity and Lorenzian model for permeability. The difference is that CRS117 material contains more magnetic inclusions compared to CR114 and thus showing the highest value of imaginary part of permeability.

One very important fact is that permeability of magnetic materials (in the range from 0 up to GHz frequencies) can contain one or multiple dispersion areas ([9]), ([10]). Below 20GHz we did not perform scattering measurements, so we have no data to be used in the fitting procedure. Furthemore, we do not have any information about materials that we examine so we cannot be completely sure of the correctness of the reconstructed model for permeability behavior at frequencies below 20GHz. For that reason, retrieved permittivity and permeability data should be used in the frequency range from 20 to 140GHz, ie. that adopted in the fitting procedure.

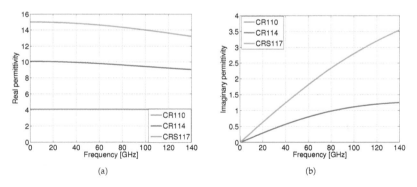

Figure 9. Real and imaginary parts of retrived permittivity of Eccosorb samples.

5.4. Extracted permittivity of epoxies Stycast W19 and Stycast 2850 FT

Commercially available epoxies Stycast W19 ans Stycast 2850FT can be used as matrices in the synthesis of microwave absorbing materials. Carbonyl iron or steel particles can be used as

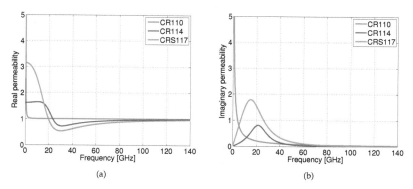

Figure 10. Real and imaginary parts of retrived permeability of Eccosorb samples.

fillers. The differences between two epoxies are in viscosity, dielectric and thermal properties. Stycast 2850FT epoxy is loaded with alumina for the higher thermal conductivity. Alumina loading makes Stycast 2850FT epoxy more viscose than Stycast W19, thus implying lower filling fraction of possible absorbing particles. Stycast W19 exhibits lower viscosity and lower thermal conductivity than Stycast 2850FT. In terms of dielectric properties, Stycast 2850FT exhibits higher real permittivity and losses than Stycast W19.

Mentioned epoxies are dielectrics and have one relaxation frequency at microwave frequencies. In order to extract frequency dependent permittivity, we model it with simple Debye model (Eq. (10)). With the same procedure described in section 5.2, we extract high frequency permittivity model of both Stycast W19 and Stycast 2850FT.

Figure 11 (a) represents measured and fitted S_{21} of Stycast 2850FT in W band. Figure 11 (b) represents measured S_{21} of Stycast 2850FT in Ka band together with simulations performed with extracted permittivity model from W band. Good agreement is achieved. Figure 12

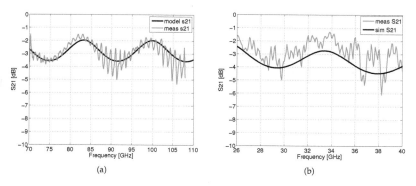

Figure 11. Measured and fitted S_{21} of Stycast 2850FT epoxy in W band (a) and measured and simulated S_{21} in Ka band (b).

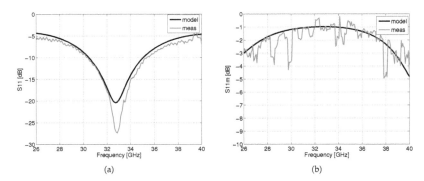

(a)   (b)

Figure 12. Measured and simulated reflection coefficient of Stycast 2850FT epoxy samples without (a) and with (b) metal backing. Good agreement between measurements and simulations is obtained which validate extracted permittivity model.

represents measured and simulated reflection of Stycast 2850FT with and without metal backing in Ka band. There is a good agreement between measured and simulated data which is proof for correctness of the extracted model. Figure 13 represents extracted real and imaginary frequency dependent permittivity of examined epoxies. Conclusion is that in the case of dielectric materials whose permittivity model exhibits one relaxation frequency in microwave frequencies, permittivity model extracted from the fitting at high frequencies is also valid at low frequencies.

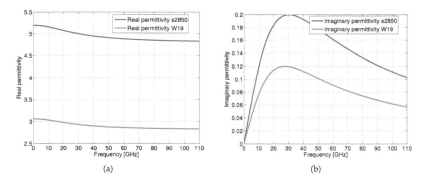

(a)   (b)

Figure 13. Real and imaginary permittivity extracted from the fitting for both Stycast W19 and Stycast 2850FT.

6. Conclusions

In this Chapter, we described a method for the extraction of frequency dependent permittivity and permeability parameters of magnetically loaded absorbing materials from free space transmission measurements. Our approach can be applied to noisy data and do not need any parameter to be known in advance. The starting assumption is based on the consideration of a two composite material (dielectric matrix and magnetic particles). According to ([8]) about models that represent composite materials, dielectric property of our samples was modeled with simple Debye relaxation model, while complex permeability was modeled with Lorenzian resonant model. Important thing was that we restored first permittivity models of the samples by fitting at high frequencies where permeability is constant and equal 1. After that step we did fitting at low frequencies to extract permeability model. The proposed method is also suitable for permittivity extraction of dielectric materials in those situations where no *a priori* information is known about material, except that material is two composite, homogeneous and isotropic. Also, for the first time we presented extracted complex and frequency dependent values of permittivities and permeabilities of Eccosorb absorbing materials (CR110, CR114 and CRS117) in the frequency range from 22 to 140 GHz. Since below 22 GHz we did not perform scattering parameters measurements and fitting, we can not say if the extracted models are also valid in that region. Future work will include investigation of permittivity and permeability frequency dependence at low frequencies (from 0 up to 22 GHz) and from 150 to 650 GHz.

Author details

Irena Zivkovic and Axel Murk
Institute of Applied Physics, University of Bern, Switzerland

7. References

[1] Bunget, I. (1984). *Physics of solid dielectrics*, Materials science monographs 19, Elsevier.

[2] Choi, H. D. et al. (1998). Frequency Dispersion Characteristics of the Complex Permittivity of the Epoxy Carbon Black Composites, *Journal of Applied Polymer Science* Vol.67.

[3] Jarvis, J. B. (1990). Transmission Reflection and Short Circuit Line Permittivity Measurements, *NIST* Technical Note 1341.

[4] Jarvis, J. B. et al. (1992). A non-linear least squares solution with causality constraints applied to transmission line permittivity and permeability determination, *IEEE Transactions on Instrumentation and Measurements* Vol. 41.

[5] Jarvis, J. B. et al. (2005). Measuring the permittivity and permeability of lossy materials: Solids, Liquids, Building Material and Negative-Index Materials, *NIST* Technical Note 1536.

[6] Kittel, C. (1946). Theory of the dispersion of magnetic permeability in ferromagnetic materials at microwave frequencies, *Physical Review* Vol.70.

[7] Rado, G. T., Wright, R. W.& Emerson, W. H. (1950). Ferromagnetism at very high frequencies. Two mechanisms of dispersion in a ferrite, *Physical Review* Vol. 80.

[8] Sihvola, A. (1999). *Electromagnetic Mixing Formulas and Applications*, The Institution of Electrical Engineers, London, UK.

[9] Zhuravlev, V. A. &Suslyaev V. I. (2006a). Physics of magnetic phenomena analysis and correction of the magnetic permeability spectra of Ba3Co2Fe24O41 hexaferrite by using Kramers-Kronig relations, *Russian Physics Journal* Vol. 49: No. 8.

[10] Zhuravlev, V. A. &Suslyaev V. I. (2006b). Analysis of the microwave magnetic permeability spectra of ferrites with hexagonal structure, *Russian Physics Journal* Vol. 49: No. 9.

Microwave Characterization of Biological Tissues

Chapter 7

The Age-Dependence of Microwave Dielectric Parameters of Biological Tissues

Mimoza Ibrani, Luan Ahma and Enver Hamiti

Additional information is available at the end of the chapter

1. Introduction

Today's children are being exposed to electromagnetic fields even before they come to this world, while they are growing up in an environment polluted, amongst others, by the dense microwave electromagnetic flux.

Daily exposure, both indoor and outdoor, of younger generations to microwave electromagnetic fields is being followed with the raised concerns regarding possible biological and health effects induced as a result of exposure. Therefore there are many published scientific papers, ongoing research projects and awarness raising campaigns aiming to inform general public and other stakeholders about safety of microwave exposure.

In order to ensure the public safety, based on research evidence, the relevant authorities have developed and announced guidelines and limits for exposure to electromagnetic fields.

Even though there are recommendations and exposure standards and limits given at international level as ICNIRP [1] and IEEE [2], few countries have set even more rigorous country specific exposure limits in comparison with international ones, in terms of setting precautionary measures.

International safety standards and guidelines on exposure limits to electromagnetic fields have been developed based on research evidence for adults, and even though each of them include a specific safety margin it should be confirmed they remain valid for children as well.

Children have longer life time exposure to microwaves in comparison with adults since they are exposed to microwave electromagnetic fields at earlier age in comparison with adults, so the cumulative exposure effect is not to be neglected.

At ICNIRP standards [1] two classes of guidance are presented: 1) Basic restrictions: Restrictions on exposure to time-varying electric, magnetic, and electromagnetic fields that are based directly on established health effects. Depending upon the frequency of the field, the physical quantities used to specify these restrictions are current density (J), specific energy absorption rate (SAR), and power density (S) .Only power density in air, outside the body, can be readily measured in exposed individuals.; 2) Reference levels: These levels are provided for practical exposure assessment purposes to determine whether the basic restrictions are likely to be exceeded. Some reference levels are derived from relevant basic restrictions using measurement and/or computational techniques, address perception and adverse indirect effects of exposure to electromagnetic fields. The derived quantities are electric field strength (E), magnetic field strength (H), magnetic flux density (B), power density (S), and currents flowing through the limbs.

At microwave frequencies the basic restrictions are given in terms of SAR- parameter used to assess absorption of electromagnetic energy by biological tissue.

The SAR, time rate of RF energy absorbed per unit mass of body biological tissue, is given as:

$$SAR = \frac{(\sigma + w\varepsilon_0 \varepsilon_r^{''})E_i^2}{\rho} \quad (W/kg) \tag{1}$$

Where

σ - Conductivity of biological tissue,

w – Angular frequency,

$\varepsilon_r^{''}$ - Imaginary part of complex permittivity,

ε_0 - Permittivity of free space,

ρ -tissue density,

E_i - induced value of electrical field as a result of exposure to external field.

As noticed from relation (1) the SAR depends directly from: induced intensity of electrical field, density of biological tissue and electromagnetic properties of biological tissues at given point. Therefore, a first step in any analysis related with SAR determination is the derivation of electromagnetic properties of biological tissues.

Looking in the other aspect the electromagnetic parameters of human biological tissues are considered very important in the wide range of biomedical applications as reported on [3] such as functional electrical stimulation, diagnosis and treatment of various physiological conditions with weak electric currents, radio-frequency hyperthermia, electrocardiography etc. On a more fundamental level, knowledge of these electrical properties can lead to an understanding of the underlying basic biological processes, for example: one of the first demonstrations of the existence of the cell membrane was based on dielectric studies on cell suspensions.

As for all mediums, the constitutive electromagnetic parameters of biological tissues are the electric permittivity ε , the magnetic permeability μ and the electric conductivity σ. These parameters describe the interaction of external electromagnetic field with medium and determine the pathways of current flow through the human body.

The conductivity of a medium may be considered as a measure of the ability of its charge to be transported throughout its volume by an applied electric field. Similarly, permittivity is a measure of the ability of its dipoles to rotate or its charge to be stored by an applied external field [3].

For almost all mediums the electromagnetic parameters vary with the frequency of the applied signal. Such a variation is called dispersion. Biological tissues exhibit several different dispersions over a wide range of frequencies, as shown in Figure 1, and described in detail in [3]. Dispersion is characterized with the relaxation time or equivalently with relaxation frequency and can be described in terms of the orientation of dipoles and the motion of the charge carriers.

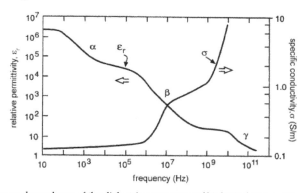

Figure 1. Frequency dependence of the dielectric parameters of biological tissues [3]

Biological tissues may be considered as materials with $\mu = \mu_0$ thus it is of interest derivation of two others parameters: permittivity and conductivity known as dielectric parameters of tissues.

Dielectric parameters of tissues are functions of frequency, but they also depend on temperature and tissue organic composition.

A few studies have presented the variation of permittivity and conductivity as a function of tissue age, triggering scientific community to explore about possible difference of microwave energy absorption between ages and finding more sensitive age- groups to electromagnetic fields exposure.

The age –dependence of dielectric properties of biological tissues mostly relies on the fact that permittivity and conductivity may be expressed as a function of tissue water content (TBW- Total Body Water Content) that is function of age, respectively decreases with age.

The TBW also differs at people of the same age while it is a function of gender as well.

The newborns have higher water content in comparison with fat while biggest part of their mass is composed by visceral organs. Beside water content also development of different organs with age is to be considered.

Therefore there are arguments to elaborate the age-dependence of dielectric properties of biological tissue, and assign different values for different age-groups, especially pointing out the difference between the children, adults and elderly people dielectric characteristics.

Up to date, based on our knowledge, there is no database with permittivity and conductivity of children biological tissues.

As a consequence, in many electromagnetic dosimetry research studies the dielectric properties of biological tissues for adults are used due to the lack of information related to children.

The age-dependence of dielectrical parameters of human biological tissues is still debatable at scientific forums and studies are mainly focused on two points:

- Finding a relation that will confirm possible age-dependence of dielectric properties of human biological tissues, respectively answering the question:
  Are the permittivity and the conductivity of human biological tissues a function of age? and
- What impact will have the age-dependent dielectric properties on induced SAR values as a result of exposure to incident electromagnetic fields, respectively in dosimetric point of view what will be the difference of microwave energy absorption by different age biological tissue for the same exposure conditions?

For determination of children biological tissues dielectric parameters, and comparison of them with adult dielectric parameters afterward, based on ongoing research, there are two major possibilities:

- *In vivo* measurement of dielectric characteristics of animal biological tissues at different animal age aiming to confirm the age dependence of dielectric properties of animal tissues, where the main raised issue is the extrapolation of obtained results from animal to human biological tissues, respectively finding a correlation of dielectric characteristics of animal and human at different ages.
  Up to date, based on our knowledge, there is no systematic scientific confirmed extrapolation methodology.
- Derivation of empiric formulas for estimating dielectric parameters of human biological tissues as a function of age.
- Possibility for derivation of permittivity and conductivity for each age via mathematical expressions.

In the chapter, the above mentioned methods will be considered, and the main studies on the field will be discussed, by outlining also their advantages and disadvantages from our point of view.

It has to be mentioned that for few external biological tissues, as for the skin as an example, there is a possibility of *in vivo* measurement for determination of dielectric properties. In a report derived by a research project [4] the study with volunteers has been conducted presenting the results for permittivity and conductivity as obtained with in vivo measurements including a detailed analysis of possible measurement uncertainities.

2. Dielectric parameters of human biological tissues derived from their correlation with animal biological tissues

For ethical reasons and due to the complexity of *in vivo* research with human biological tissues, especially for children, few researchers have conducted experiments with animal biological tissues tending to confirm the age-dependence of dielectric parameters of biological tissues.

One of the most systematic and cited studies dealing with dielectric properties of biological tissues remains report [5] based on which the database and software applet is developed presenting the values of permittivity , conductivity and the other dielectric characteristics, for main biological tissues at microwave frequencies.

The presented results of permittivity and conductivity of human biological tissues have been obtained by experimental techniques on three types of materials: excised animal tissues, human autopsy materials and in vivo human skin and tongue, whose details are given in [5].

The comparison between dielectric characteristics of human and animal biological tissues and between animal species is given on Figures 2, 3 and 4.

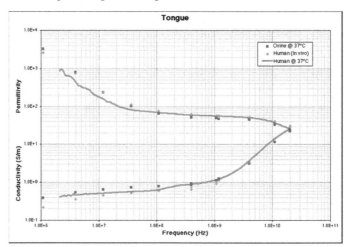

Figure 2. Comparison between the dielectric properties of tongue muscle from animal and human samples [5]

The difference between dielectric parameters of animal and human is not systematic, thus it has to be noted that the difference between dielectric parameters of the same specie may sometimes exceed the difference between the different species.

These differences make the method of extrapolation of dielectric parameters from one specie to the other not sustainable at desired extent.

Nevertheless, from the above figures, the similarity of dielectric parameters variation trend between the animals and human biological tissues can be observed.

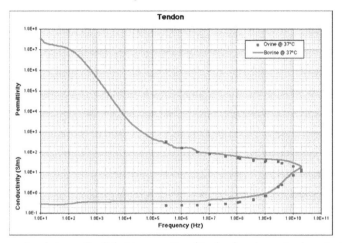

Figure 3. Comparison between the dielectric properties of tendon from two animal species [5]

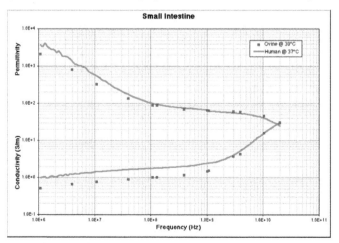

Figure 4. Comparison between the dielectric properties of small intestine tissue from animal and human samples [5]

Even the study gives detailed results and deep description of research methodologies, providing broad based consensus results on the subject, the age-dependency of dielectric characteristics of biological tissues is not included in software application and the user can not obtain the values of permittivity and conductivity as a function of age.

Few studies have reported variation of dielectric parameters with age mostly for animal biological tissues.

Thurai et al. 1984[6] and 1985[7] reported a change on values of dielectric parameters as a function of age for brain of animal tissue.

Peyman et al. [8,9] presented the values of experimental measurements of dielectric parameters of rats aged 10-70 days, thus confirming the age-dependence of permittivity and conductivity of the animal biological tissues. In the study, the dielectric properties of ten rat tissues at six different ages were measured at 37 degrees celzius in the frequency range between 130 MHz and 10 GHz, using an open-ended coaxial probe and a computer controlled network analyser. The results showed a general decrease of the dielectric properties with age, with a trend being more apparent for skull, skin and brain tissues and less noticeable for abdominal tissues. The variation in the dielectric properties with age reported in the study is due to the changes in the water content and the organic composition of tissues. The results presented in study have given some insight to research community for possible differences and comparative assessment of exposure between children and adults.

In the review article [10], experimental results are presented on the dielectric properties of tissue samples of 10, 30, and 70 days-old Wistar strain rats in the frequency range 300 KHz– 300 MHz. The study concluded that at frequencies higher than 100 MHz, where the γ dispersion is dominant, the permittivity and the conductivity increase monotonically with decreasing age. At lower frequencies, the site of the β dispersion, a change in the frequency dependence of the dielectric parameters is observed and is mostly evident in the spectra for brain and skin. According to this research, it is attributed to changes in the tissue structure.

In the same study [10] age-related dielectric data were incorporated in a numerical plane wave exposure dosimetry applied on anatomically heterogeneous rat models with body sizes corresponding to the ages of 10, 30, and 70 days at a number of spot frequencies from 27 to 2000 MHz. The obtained results showed that the variation in the dielectric properties affect the whole body SAR by less than 5% with the most conservative value (highest SAR) obtained when 70 day properties were used.

Schmid and Uberbacher [11] have presented the results of experimental measurements confirming that conductivity and permittivity of younger animal tissues were for 15-22 %, respectively 12-15 % , higher than the ones of older animal tissues.

Peyman et. al. [12, 13] also conducted measurements of dielectric parameters of pig biological tissues at different ages, re-confirming the age- dependence of dielectric characteristics of animal tissues, and in the same study the correlation between the animal dielectric parameters and human dielectric parameters is given.

The dielectric properties of pig cerebrospinal tissues were measured *in vivo* and *in vitro*, in the frequency range of 50 MHz–20 GHz. The total combined measurement uncertainty was calculated at each frequency point and is reported over representative frequency regions. Comparisons were made for each tissue between the two sets of data and with the literature of the past decade. The *in vitro* study was extended to include tissue from pigs weighing approximately 10, 50 and 250 kg to confirm the issue of the variation of dielectric properties with age.

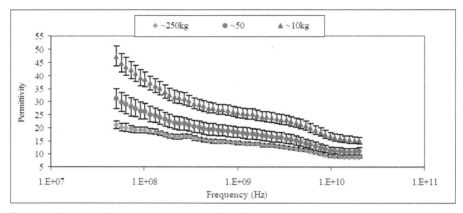

Figure 5. The measured permittivity of the long bone for different age pigs [4]

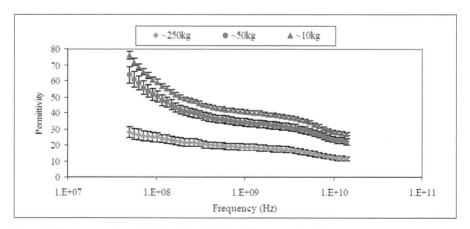

Figure 6. The measured permittivity of the skull for different age pigs [4]

As reported on [4] the age related effects were more noted for some tissues types. Results of porcine study showed that at tissues: white matter, dura, fat, skin, skull and spinal cord the age related effects were noticed, while at tissues as: tongue, cornea and grey matter no age related effects were noticed.

For some tissues at microwave frequencies there was no significant change between results obtained from *in vitro* and *in vivo* measurements.

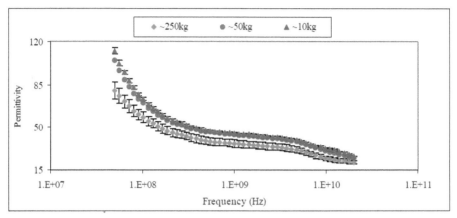

Figure 7. The measured permittivity of the skin for different age pigs [4]

Recalling the study of Peyman at al . [13], based on discussions with numbers of vets, the studies on pig and human growth curves with markers along curves (e.g., sexual maturity etc.), a partial extrapolation of results to humans is reported.

The study concluded that it was more straightforward to correlate the end points, i.e., the piglets to be equivalent to very small children and mature pigs equivalent to adults, whereas the ages in between are more difficult to correlate.

The study reports : 10 kg pigs, which are less than 30 days old, are very young animals and far from puberty age; - it may be assumed that dielectric properties of these piglets correspond to those of human children of age between 1-4 years. The 50 kg pigs , which are about 100 days old , are not still fully matured in terms of sexual activity, and therefore are assumed to be equivalent to 11-13 years old human. Finally, at 250 kg, pigs are fully matured animals and are therefore considered equivalent to human adults.

The results obtained with extrapolation were then used to calculate the SAR values in children of age 3 and 7 years when they are exposed to microwave induced by walkie–talkie devices.

Even though a first steps were made to find a correlation between dielectric parameters of animal and human at different age, still there is no systematic scientific-based general correlation formula or extrapolation methodology that would propose the extrapolation of obtained results of animal dielectric parameters to humans.

3. Age-dependent dielectric parameters of human biological tissues derived with analytical formula

The complex permittivity of any tissue, including here the biological tissues, the parameter that contains both: tissues relative permittivity and conductivity, is given by the formula:

$$\underline{\varepsilon}_r = \varepsilon'_r - j\varepsilon''_r = \varepsilon_r - j\frac{\sigma}{\omega\varepsilon_0} \tag{2}$$

where:

$j = \sqrt{-1}$

- Relative permittivity.

If on relation is added

τ - the relaxation time of the biological tissue given by the folowing expression:

$$\tau = \frac{\varepsilon_0}{\sigma}\varepsilon_r \tag{3}$$

By substituting the relaxation constant into the complex permittivity formula (2), the following expression is obtained:

$$\underline{\varepsilon}_r = \varepsilon_r\left(1 - j\frac{1}{\omega\tau}\right) \tag{4}$$

The plot of the relation between the imaginary and the real part of permittivity at frequency 900 MHz, for few aging rats biological tissues as: the brain, the muscle, the salivary gland, the skin and the skull, based on results presented on [8], confirms that ratio between the imaginary and real part of permittivity is almost constant for different biological tissues as described on [14].

Therefore, it may be concluded that the relaxation constant is almost independent from age and its value can be taken statistically as a mean derived from different ages.

Under the above mentioned assumption, either relative permittivity or conductivity of biological tissue has to be known and the other can be calculated using relation (3).

Referring to mathematical expression (4) the only parameter that may change with age is tissue relative permittivity ε_r.

The dielectric parameters of tissue depend on tissue composition.

It is known that human biological tissues are dominantly constituted by water content, therefore if water content changes with age it is to be expected that dielectric parameters change with age as a result of TBW change, and this has to be put in a formula expressing the function of relative permittivity from age of tissue, respectively TBW, approach proposed by Wang at al. [14] on 2006.

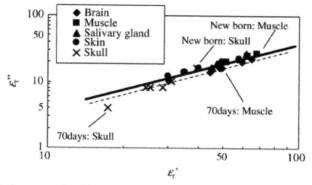

Figure 8. Ratio between real and imaginary part of permittivity of biological tissues [14]

3.1. Empirical derivation of dielectric parameters of human tissues as a function of age

The relative permittivity based on Leichtencker's logarithmic mixture law [15] for medium consisted from N composites may be expressed as:

$$\varepsilon = \prod_{n=1}^{N} \varepsilon_n^{\alpha_n} \tag{5}$$

ε_n - relative permittivity of n composite

α_n - hydration rate of n composite

with sum of all α equal to one.

For medium consisted from two composites the relative permittivity is given as:

$$\varepsilon = \varepsilon_1^{\alpha} * \varepsilon_2^{1-\alpha} \tag{6}$$

As noted on [15] Leichtencker mixture formula is symmetric with respect to its constitutes thus:

$$\frac{\varepsilon}{\varepsilon_2} = (\frac{\varepsilon_1}{\varepsilon_2})^{1-\alpha} \text{ can be written as } \frac{\varepsilon}{\varepsilon_1} = (\frac{\varepsilon_2}{\varepsilon_1})^{\alpha} \tag{7}$$

Even though formula (6) served as a starting point to Wang et al. [14], 2006, it has been regarded for a long time as semi-empirical formula without theoretical justification. Recently, a 2010 study [16] has confirmed that formula (6) can be derived by applying

Maxwell's equations and the principle of charge conservation to a mixture for which the spatial distribution of shapes and orientations of the components is randomly distributed.

According to Wang [14] the biological tissues may be considered as given by two composites: water and organ specific material, therefore the relative permittivity of human biological tissues may be given as:

$$\varepsilon_r = \varepsilon_{rw}^{\alpha} \cdot \varepsilon_{rt}^{1-\alpha} \tag{8}$$

ε_{rw} - relative permittivity of water,
ε_{rt} - organ specific relative permittivity
α – tissue hydration rate.

The tissue hydration rate, as a function of TBW and mass density ϱ, is expressed by the relation:

$$. \alpha = \rho \cdot TBW \tag{9}$$

Even though the other compounds, organ specific ones, vary with age, for simplification is considered that only hydration rate is age dependent.

The mass density is not considered as a function of age, therefore the only parameter that reflects change with age is TBW, while TBW variation with age has to be quantified in order to derive expression/formula that gives permittivity as a function of age.

Wang at al.[14] has proposed the fitting function for TBW which tends to fit values of TBW as a function of human age, values published by Altman and Dittmer on 1974 [17].

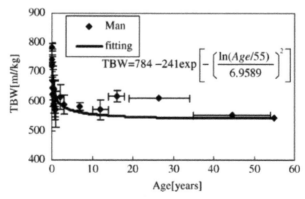

Figure 9. Fitting function for TBW [14]

After few mathematical operations the formula for calculating complex tissue permittivity is extracted, as a function of tissue hydration rate, adult tissue permittivity and adult tissue hydration rate:

$$\underline{\varepsilon}_r = \varepsilon_{rw}^{\frac{\alpha-\alpha_A}{1-\alpha_A}} \cdot \varepsilon_{rA}^{\frac{1-\alpha}{1-\alpha_A}} \left(1 - j\frac{1}{\omega\tau}\right) \tag{10}$$

α - child biological tissue hydration rate

α_A - adult biological tissue hydration rate

ε_{rA} - relative permittivity of adult biological tissue

Even though it is a very good approach on finding dielectric parameters of humans at different ages, the following remarks need to be raised up: Fitting function values deviate considerably from experimental measurements for some ages and according to figure the TBW varies at great extent under 3 years old but becomes insignificant for ages over 3 years old.

The very important aspect is the formula validation. Wang at al. [14] report validation of formula by checking validity using Peyman at al. [8] measured data for rats. The difference of results obtained with formula and measure data for rats were within $\pm 20\%$ for all the tissues in the young rats, thus empirical formula proposed by Wang is therefore the reasonable representation of age-dependent tissue dielectric properties.

3.2. Dielectric parameters of 3 months-13 year's old boys biological tissues

Wang et al. [14], as described on previous paragraph, derived an empirical formula for the calculation of age dependency of permittivity and conductivity of human biological tissues as a function of the total body water content (TBW), where TBW values were taken from Altman and Dittmer [17] measurements of year 1974.

As reported in [18], secular trends in the nutritional status of infants and children alter the relation between age or weight and TBW, thus equations proposed in previous decades overestimated TBW in all age groups, including children.

TBW at humans has changed from year 1974 and especially in children. There are lately revised formulas that give relation of TBW as a function of child height and weight.

Therefore, in [19] we propose the formulas for calculation of permittivity and conductivity of children biological tissues as a function of children weight and height, respectively age, whereas child TBW values are calculated with formula presented on [18].

Children biological tissues are composed mainly of water, thus for simplification reason they may be considered as given by two composites, water and organ specific tissue. Even though concentration of other compounds, organ specific ones, may also vary with age, for simplification purpose we assume that organ specific part is not age dependent, while main part of biological tissue TBW changes as a function of age.

For the child biological tissues, permittivity can be expressed as:

$$\varepsilon_{rch} = \varepsilon_{rw}^{\alpha_{ch}} \cdot \varepsilon_{r0}^{1-\alpha_{ch}}$$

(11)

α_{ch} - Child biological tissue hydration rate

Following mathematical operations, as described at Wang et al. [14], permittivity as a function of hydration rates, respectively child and adult TBW, can be formulated as:

$$\varepsilon_{rch} = \varepsilon_{rw}^{\frac{\alpha_{ch}-\alpha_A}{1-\alpha_A}} \cdot \varepsilon_{rA}^{\frac{1-\alpha_{ch}}{1-\alpha_A}}$$

(12)

Lately, the Mellits and Cheeks formula for the determination of child TBW has been reformulated. According to [18], child TBW in liters for age 3-month-old to 13-year-old boys can be expressed as:

$$TBW = 0.0846 \left(Height \cdot Weight \right)^{0.65}$$

(13)

For girls for age 3-month-old to 13-year-old the TBW is expressed as:

$$TBW = 0.0846 * 0.95 \left(Height \cdot Weight \right)^{0.65}$$

(14)

When replacing TBW for boys on above equations and after few approximations we derived formulas (15) and (18) that were presented in [19].

In order to derive formulas we made some approximations. For tissues, we considered mass density 1,071 g/ml that might be taken as an average of mass density for most tissues, but not all of them. Some organs, such as lungs, do have mass density that differs significantly, up to 3 times, from the taken average value. For such tissues the proposed formula needs to be adopted.

In order to find approximate adult hydration rate, and take it as a constant in formula, we assume adult weight as 75 kg and TBW as 41.9 liters. Based on measurements conducted on [20], 41.9 liters corresponds to a man who is 20-29 years old.

For relative permittivity of water the value of 74.3 at 37 Celsius is taken [21].

Presented formulas express the relative permittivity of child biological tissue as a function of child height and weight and a function of an adult relative permittivity.

The formulas are considered valid if we refer other values of adult TBW and weight that do not reflect significant change of adult hydration rate.

$$\varepsilon_{r_{ch}}(X) = 2.616^{(X-6.63)} * \varepsilon_{r_A}^{2.4813(1-0.09X)}$$

(15)

The variable X is introduced to simplify the formula in terms of its dependence on child height and weight:

$$X = Height^{0.65} * Weight^{-0.35} \tag{16}$$

Referring to relation (3) , the conductivity of tissue as function of relative permittivity may be expressed as:

$$\sigma_{ch}(X) = \frac{10^{-9}}{36\pi\tau} \varepsilon_{r_{ch}}(X) \tag{17}$$

Replacing relation (15) on (17) the following relation for conductivity is obtained:

$$\sigma_{ch}(X) = \frac{10^{-9}}{36\pi\tau} \varepsilon_{r_{ch}}(X) = 2.616^{(X-6.63)} * \varepsilon_{r_A}^{2.4813(1-0.09X)} \tag{18}$$

With the formula (15) and (16) the values of permittivity of child bio tissues are calculated and presented in Figures 10-11.

Adult tissue relative permittivity values were taken from [5] while child height and weight were taken from World Health Organization growing charts, as 50th growing percentiles [22].

The advantage of the proposed approach is the calculation of dielectric parameters, conductivity and permittivity, for biological tissues of children of specific height and weight.

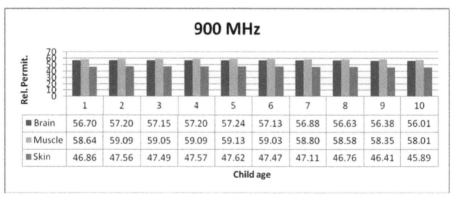

	1	2	3	4	5	6	7	8	9	10
■ Brain	56.70	57.20	57.15	57.20	57.24	57.13	56.88	56.63	56.38	56.01
▣ Muscle	58.64	59.09	59.05	59.09	59.13	59.03	58.80	58.58	58.35	58.01
■ Skin	46.86	47.56	47.49	47.57	47.62	47.47	47.11	46.76	46.41	45.89

Figure 10. Children's relative permittivity at 900 MHz

The results are presented for brain, muscle and skin for children aged 1-10 years old, boys, for frequencies 900 MHz, 1800 MHz and 2.4 GHz.

From Figures 8-10, it can be easily extrapolated the difference between biological tissue dielectric parameters of end ages, 1-year-old and 10-year-old children, but it can be also made a comparative analysis between the permittivity of the same child age but different biological tissue.

Even though the values at different frequencies change, the similar variation trend of tissue permittivity as a function of age is noticed for all microwave frequencies

If we compare obtained permittivity values among different tissues, we notice that highest values are obtained for the muscle and the lowest ones for the skin.

It should be mentioned that height and weight were taken as 50th growing percentiles in relation age- (height and weight) while the methodology for extracting values for other growing percentiles remains unchanged.

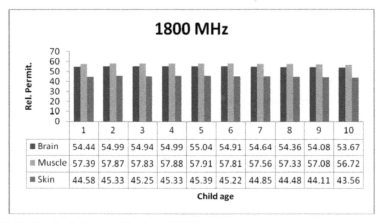

Figure 11. Children's relative permittivity at 1800 MHz

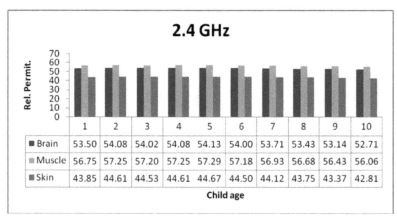

Figure 12. Children's relative permittivity at 2400 MHz

After obtaining results for children biological tissues dielectric parameters, the comparison between obtained values and adult bio tissues dielectric parameters can be conducted.

In Figure 13 is presented only one example of comparative analysis, namely the differences on muscle permittivity for frequency 900 MHz.

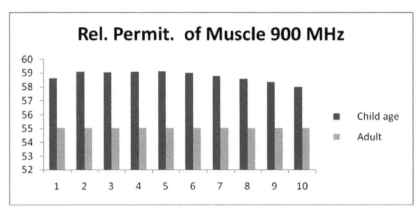

Figure 13. Child vs. adult relative permittivity at 900 MHz

It should be pointed out that all obtained values of children dielectric parameters are results of proposed theoretical approach which needs experimental validation; therefore the proposed formulas need experimental validation.

The aim of the study was to confirm age-dependence of dielectric parameters and to verify that child dielectric parameters differ from adults and they should not be taken as the same in the dosimetric research, as many research studies have done so far.

The knowledge of dielectric properties for humans at different age is important as well to create body-simulant materials that will be used for electromagnetic measurement applications. These materials should have dielectric properties that mimic those of human tissues [23].

The human body is composed by multilayered tissues, dielectric parameters of each of them vary with age, some more some less. For simplification purpose multilayered tissues can be simulated by one homogenous liquid whose dielectric properties match those of the tissue that is of most interest [24].

Beside comparative analysis between children and adult dielectric parameters of biological tissues, the very limited comparative analysis between child tissues permittivity obtained by different studies was conducted.

As noticed in the Figures 14-15, there is a good agreement between values obtained by different studies.

Nevertheless the comparative research should be more inclusive conducted for more tissue in order to come to exhaustive conclusions.

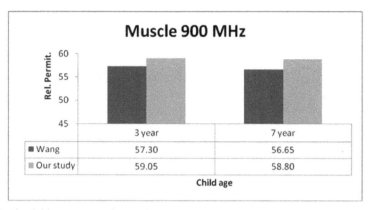

Figure 14. The child permittivity obtained by two studies

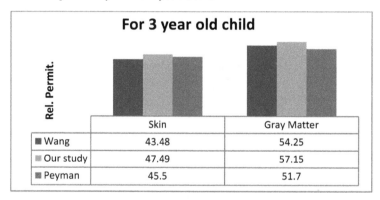

Figure 15. The 3-year old child permittivity obtained by three studies

4. Conclusions

The confirmation of age-dependence of dielectric parameters of human biological tissues potentially may contribute on issue whether children should be considered a sensitive group in comparison to adults regarding exposure to microwave electromagnetic fields, issue that still remains debatable at scientific forums.

Experimental measurement *in vivo* at most human biological tissues is impossible, thus alternative methodologies have to be considered, few of them mentioned on this chapter.

Extrapolation of obtained experimental results of dielectric parameters of animal bio tissues to humans is considered debatable since animal tissues differ significantly from human tissues, and a detailed and all –inclusive study of animal and human growing curves along with markers along curves should be performed to propose correlation between animal and human dielectric parameters.

Thus, presenting the empiric formulas for estimating permittivity and conductivity of human biological tissues as a function of age is to be considered as a reasonable solution for the issue even though the each proposed theoretical approach needs experimental validation, which is very limited due to available measuring techniques.

Since biological tissues are dominantly composed of water content, their dielectric characterization is a function of TBW- a parameter that changes with age. This has triggered researchers to propose formulas for estimating permittivity and conductivity of human biological tissues as a function of age, respectively TBW.

The results presented in the chapter, with all covered methods, confirm that permittivity and conductivity of human tissues are age-dependent, presenting also change trend over age.

The limited comparative analysis shows a satisfactory agreement between results obtained with different approaches.

Nevertheless it should be pointed that results presented with theoretical approach are not experimentally validated, so they should be taken with reserve by the scientific community. As a matter of fact, the aim was to confirm the variation trend of permittivity and conductivity with age and not to present precise values of microwave dielectric parameters of human tissues for each human age.

Experimental validation of theoretical approaches for derivation of dielectric parameters of human tissues as a function of age, detailed comparative analysis of obtained results by different studies, assessment of difference between dielectric parameters of males and females at each age, are just a few of issues that remain open for further research.

Appendix

Mimoza Ibrani, Luan Ahma and Enver Hamiti
Faculty of Electrical and Computer Engineering, University of Prishtina, Republic of Kosova

5. References

[1] ICNIRP, Guidelines for limiting exposure to time-varying electric, magnetic, and electromagnetic fields (up to 300 GHz), "*Health Phys.*, Vol. 74, No. 4, 1998.

[2] *IEEE Standards for Safety Levels with Respect to Human Exposure to Radio Frequency Electromagnetic Fields, 3 kHz to 300 GHz*, IEEE Standard C95.1, 2005.

[3] Electric properties of tissues by D Miklavcic, N Pavselj, F X Hart Wiley encyclopedia of biomedical engineering (2006) Volume: 209, Issue: 922834401, Publisher: Wiley Online Library, Pages: 1-12, ISSN: 00247790, DOI: 10.1002/9780471740360.ebs0403, PubMed: 13981120

[4] Dielectric Properties of Tissues at Microwave Frequencies, A. Peyman, S. Holden and C. Gabriel, RUM 3 report,
http://www.mthr.org.uk/research_projects/documents/Rum3FinalReport.pdf

[5] *Compilation of the dielectric properties of body tissues at RF and microwave frequencies,* Gabriel C and Gabriel S, 1996, *Technical report AL/OE-TR-1996-0037*

[6]   Thurai M, Goodridge VD, Sheppard RJ and Grant EH (1984) Variation with age of the dielectric properties of mouse brain cerebellum. *Phys Med Biol. 29*, 1133-1136.

[7]   Thurai M, Steel MC, Goodridge VD, Sheppard RJ and Grant EH (1985) Dielectric properties of developing rabbit brain at 37⁰ C. *Bioelectromagnetics 6*, 235-242.

[8]   Peyman A, Rezazadeh AA and Gabriel C (2001) Changes in the dielectric properties of rat tissue as a function of age at microwave frequencies. *Phys Med Biol 46*, 1617-1629.

[9]   Peyman A, Rezazadeh AA and Gabriel C (2002) Changes in the dielectric properties of rat tissue as a function of age at microwave frequencies. Corrections to Peyman et al. (2001). *Phys Med Biol 47*, 2187-2188

[10] Gabriel, C. (2005), Dielectric properties of biological tissue: Variation with age. *Bioelectromagnetics, 26: S12–S18*. doi: 10.1002/bem.

[11] Schmid G, Uberbacher R (2005) Age dependence of dielectric properties of bovine brain and ocular tissue in the frequency range of 400 MHz to 18 GHz, Phys. *Med. Biol. 50*, 4711-4720

[12] Peyman A, Holden S J , Watts S , Perrott R, Gabriel C (2007) Dielectric properties of porcine cerebrospinal tissues at microwave frequencies: in vivo, in vitro and systematic variations with age, *Phys. Med. Biol. 52* , 2229 – 2245

[13] *Variation of the dielectric properties of tissues with age: the effect on the values of SAR in children when exposed to walkie–talkie devices, A Peyman et al 2009* Phys. Med. Biol. *54 227 doi:10.1088/0031-9155/54/2/004*

[14] Wang J, Fujiwara O and Watanabe S, "Approximation of aging effect on dielectric tissue properties for SAR assessment of mobile telephones", *IEEE Transactions on Electromagnetic Compatibility,* Vol.48, No.2, 2006

[15] Lichtenecker K., " Die dielekrizitatskonstante naturlicher und kunstlicher mischkorper" , *Physikalische Zeitschrift,* vol.27,pp115-158, 1926

[16] Simpkin R., "Derivation of Leichtencker's Logarithmic Mixture Formula From Maxwell's Equations", *IEEE Transactions on Microwave Theory and Techniques,* Vol. 58, No. 3, 2010

[17] P.L. Altman and D.S. Dittmer, *Biology Data Book: Blood and Other bloody Fluids.*Washington, DC: Federation of American Societies for Experimental Biology, 1974

[18] Morgenstern B., Mahoney D. and Warady B.,"Estimating Total Body Water in Children on the Basis of Height and Weight: A Reevaluation of the Formulas of Mellits and Cheek",*J Am Soc Nephrol* 13:1884-1888, 2002

[19] M. Ibrani, L. Ahma and E. Hamiti " Derivation of electromagnetic properties of child biological tissues at radio frequencies", *Progress in Electromagnetic Research Letters,* Vol.25, 87-100, 2011

[20] Chumlea W, Guo Sh., Zeller Ch., Re N., Baumgartner R., Garry Ph., Wang J., Pierson R., Heymsfields S., and Siervogel R., "Total body water reference values and prediction equations for adults", *Kidney International, Vol. 59 (2001),* pp. 2250–2258

[21] Stogryn A.," Equations for calculating the dielectric constant of saline water", *IEEE Transactions on Microwave Theory and Techniques,* vol. MMT-19, no.8, pp. 733-736, 1971

[22] WHO growing charts, http://www.who.int/childgrowth/standards/en/

[23] Basic Introduction to Bioelectromagnetics, Cynthia Furse, Douglas Christensen and Carl Durney, CRC Press, 2009, ISBN-13: 978-1420055429

[24] Electromagnetic Fields and Radiation: Human Bioeffects and Safety, Riadh W.Y. Habash , Marcek Deker, 2002, ISBN-13: 978- 0824706777

Experimental Requirements for *in vitro* Studies Aimed to Evaluate the Biological Effects of Radiofrequency Radiation

Olga Zeni and Maria Rosaria Scarfì

Additional information is available at the end of the chapter

1. Introduction

In the last years human exposure to electromagnetic fields (EMF) in the radiofrequency (RF) range has increased rapidly becoming unavoidable. As a matter of fact, many sources of RF fields are present at home, at work, and in the environment. In addition, other sources of occupational RF field exposure include equipments such as medical devices, dielectric heaters, induction heaters, diathermy machines, plasma discharge equipment and radars. Moreover, the rapidly increased use of mobile telecommunication has lead to exposure of a large amount of the population to RF fields. It has been estimated that 4.9 billions of mobile phone subscriptions will be active by the end of 2012. Although these technologies have highly improved the quality of life, at the same time they have given rise to great concern about possible health effects of such non-ionizing radiation at low exposure levels, with particular attention to cancer risk. As a matter of fact, heating is the most widely accepted mechanism of RF radiation with biological systems, and the current guidelines of human exposure are based on thermal effects, but subtle effects due to chronic exposures cannot be excluded. The latter, non thermal effects, are hypothesized to occur in the absence of local or whole body increases in temperature, although there are no generally accepted biophysical mechanisms that could explain such effects.

Three main approaches, epidemiological studies, in vivo studies and in vitro studies, providing different and complementary information, can be followed in addressing the evaluation of biological effects induced by exposure to RF fields. The epidemiological studies aim to test, on a statistical basis, whether a causal nexus between exposure to an environmental agent and its putative health effects on the health status of the exposed subjects could exist. They use specially designed studies that try to determine statistical associations between independent (level of exposure) and dependent (health status, disease occurrence, etc.) variables by collecting data from population samples. In addition, in relation to RF-based wireless communications, there are two different exposure situations: to RF in the far field, emitted by base stations, WiFi access points, etc. and to RF in the near field, emitted by handheld devices (e.g. mobile phones). According to the World Health Organization publication on Electromagnetic Fields, Environmental Health Criteria series [1, 2], to proper address human health risk assessment, epidemiological research should allow for sufficient latency, sufficient range of exposure, including high exposure, and ability to accurately classify individuals into several exposure groups.

In vivo studies are carried out on human volunteers or animals and provide information concerning the interaction of RF radiation with living systems displaying the whole body functions, such as immune response, cardiovascular changes, and behaviour. There are obvious limitations in the exposure conditions to be tested on humans due to ethical issues, therefore most of the in vivo studies are carried out on laboratory animals (mainly rodents). However, extrapolation to humans to provide an estimation of health risk is not straightforward due to the differences in physiology and metabolism between species as well as differences in life expectancy and many other variables.

In vitro studies, carried out mainly on cell cultures or isolated tissue samples, are used extensively in toxicological investigations. This is because they can provide essential information about the potential effects of chemicals and other agents such as radiation on specific cell properties, and provide a more rapid and cost-effective approach to molecular and mechanistic studies than conventional laboratory animal models.

In the last 20 years, scientific investigations on whether RF radiation used in these technologies could have short-, medium- or long-term biological effects, and whether they could represent a health hazard to human population have largely increased, since any detectable detrimental effect of RF radiation, even small, could be very important, due to its widespread use, the large numbers of people exposed on a daily basis, and the social, economical and health impact this could have.

Recently, on May 2011, the International Agency for Research on Cancer (IARC) classified RF electromagnetic fields as possibly carcinogenic to humans (Group 2.B) [3, 4]. The IARC evaluated available literature about the carcinogenicity of RF electromagnetic fields and found the evidence to be "limited for carcinogenicity of RF-EMF, based on positive associations between glioma and acoustic neurinoma and long term exposure". The conclusion of the IARC was mainly based on the INTERPHONE epidemiological study, which found an increased risk for glioma in the highest category of heavy users (30 minutes per day over a 10 year period), although no increased risk was found at lower exposure. The evidence for other types of cancer was found to be "inadequate" [5]. In vivo and in vitro studies, carried out so far, have provided only limited support for the above mentioned classification [6], mainly due to the difficulty in comparing the results of the available studies to draw general conclusion. The difficulty often arises from the quality of the studies, in terms of study design, specific methodologies and analysis of the results.

In this chapter the focus is on in vitro studies, which are the studies providing insight into the basic mechanisms by which effects might be induced in more complex animal or human organisms. As a matter of fact, in vitro studies, carried out on tissue or cell cultures of animal or human origin, transformed or not transformed, are relatively simple and represent well described models, where the control of exposure and experimental conditions is significantly greater than in live animals or human volunteers.

In vitro studies, the most common in the evaluation of biological effects of RF radiation, are mainly aimed to investigate cellular endpoints related to cancer occurrence. Carcinogenesis is a multi-step process in which direct (genotoxic) and indirect (non-genotoxic) DNA damage is involved, as schematically depicted in figure 1. Therefore, cancer related studies can be classified as genotoxic and non-genotoxic. Genotoxic effects include DNA strand breaks, micronucleus formation, mutation and chromosomal aberration. Non-genotoxic effects refer to changes in cellular function, and include cell proliferation, oxidative metabolism, apoptosis or programmed cell death, cellular signal transduction, and gene expression (RNA and protein).

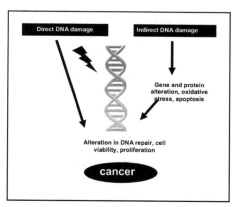

Figure 1. Schematic representation for genotoxic and non-genotoxic carcinogenesis

On the basis of the conclusions reported in the above mentioned reviews [1-3, 6], also confirmed by more recent scientific literature, the following considerations can be drawn. Results on genetic effects and other cellular endpoints, like cell proliferation and differentiation, apoptosis and cell transformation, are mainly negative, and some of the few positive findings may be attributable to a thermal insult rather than to the RF-exposure as such. As a matter of fact, some studies following replication, under more controlled conditions, failed to be confirmed by independent research groups [7]. The same is for expression of cancer-related genes (e.g., proto-oncogenes and tumor suppressor genes), and studies, carried out using powerful high-throughput screening techniques capable of examining changes in the expression of very large numbers of genes and proteins. It should be pointed out that the results achieved by high-throughput techniques need to be confirmed by quantitative methods since methodologies are not sufficiently standardized.

In order to improve the quality of in vitro studies and of the future research, high methodological quality is needed to address uncertainties in technical and biological aspects, and to determine whether the achieved results reflect "true" biological response or they are related to some unknown uncontrolled variable.

The aim of this chapter is to describe the main requirements for a well conducted in vitro study to overcome methodological limitations. Due to the inter-disciplinary nature of the bioelectromagnetic research, basically, biological and electromagnetic requirements can be identified.

2. Biological requirements

Cell cultures are mainly used as an inexhaustible source of experimental material. The use of cell cultures offers the advantage of investigating the specific interactions at cell or molecular level, and the ability to perform large series of experiments under the same conditions. Cell cultures are achieved through enzymatic or mechanical disaggregation of a tissue sample. Human and animal cell cultures are derived either from a primary explant or a cell line. A fresh isolate of cells which is cultured in vitro is called a 'primary culture' until cells are subcultured or passaged. Primary cell cultures are generally heterogeneous, with a low fraction of growing cells, but they contain a variety of cell types which are representative of the tissue. The subculture allows the propagation of the culture, which is now called a 'cell line': it appears to be more uniform, but specialized cells and functions can be lost. The greatest advantage of a cell line is the availability of a large amount of homogeneous material to be used for long periods of time due to their ability to propagate and divide into replicates. After several passages a cell line may die (finite cell line) or 'transform' to become an established or continuous cell line. Most of the continuous cell lines originate from neoplastic tissues, but several continuous cell lines are derived from normal embryonic tissue. They can be characterized and stored by freezing.

According to the Good Cell Culture Practice [8], the maintenance of high standards is fundamental to all good scientific practice, and is essential for ensuring the reproducibility, reliability, credibility, acceptance and proper application of any produced results. The standardization of cell culture is required, and is achieved by controlling the materials, such as cells and culture medium, that interact and determine the properties of the total system. However, the potential for variation can also be considered for each separate component. It is recommended that authenticated stocks of continuous cell lines are purchased from recognized national and international cell banks. Cell culture medium is a defined base solution including salts, aminoacids and sugars supplemented with cell type specific components. The more complex component is the serum that origins from a pool of donations taken from a large number of animals, thus expressing a large variability among different manufactures. Primary cell cultures require complex nutrient media, supplemented with animal serum and other non-defined components, thus primary cell culture systems are difficult to standardize. Immortalized cell lines, being able to multiply for extended periods, can be expanded and cryopreserved as cell bank deposits. They represent a more stable and reproducible system than primary cells.

A wide variety of cell types, ranging from stem cells to highly differentiated tissue specific cells, can be used. The appropriate cell model has to be used for specific experimental approaches for the proper identification of biological effect. It has to be chosen on the base of the cellular target investigated. For instance, human lymphocytes are largely employed, since they are of human origin and easy to obtain by venipuncture. They represent one of the best suited cell model for the investigation of genotoxic effects of chemical and physical agents, including EMF.

More than one endpoint have to be investigated, for each cellular target, also to balance mechanistic vs. toxicity studies. Thus, a combination of techniques, confirming and/or complementing each other, is recommended for the reliable detection of effects. For instance, to study direct DNA damage (genotoxicity), mainly cytogenetic techniques are employed, which allow to investigate the frequencies of chromosomal aberrations, sister chromatid exchanges and micronuclei whose increased level in human lymphocytes are predictive for cancer risk [9-12]. However, such assays essentially reveal severe genetic damage and could not be the optimal choice to detect most of the subtle indirect effects that may be induced by RF radiation. More sensitive techniques, like single cell gel electrophoresis (comet assay) or the detection of γ-H2AX phosphorylated histone, will help in this case. The alkaline version of the comet assay, first introduced in 1988 [13], allow to detect the combination of DNA single-strand breaks (SSBs), double-strand breaks (DSBs) and alkali-labile sites in the DNA, although the results and the interpretation of the biological relevance of the damage can be misleading if not compared with other measures of DNA damage. For the assay, cells are embedded in agarose on a microscope slide, lysed with detergent, subjected to electrophoresis and stained with a fluorescent DNA-binding dye and subsequently observed by fluorescence microscopy. Negatively charged loops/fragments of DNA migrate out of the nuclei forming a tail in the direction of the anode, giving the nuclei the appearance of a comet [14]. In Figure 2 human lymphocytes processed for comet assay are shown.

The phosphorylation of the histone variant H2AX to γ-H2AX–containing nucleosomes [15] is one of the earliest marks of a DNA double-strand break in eukaryotes. γ-H2AX is essential for the efficient recognition and/or repair of DNA double-strand breaks and many molecules, often thousands, of H2AX become rapidly phosphorylated at the site of each nascent double-strand break. It was shown that this simple method was suitable to monitor response to radiation or other DNA-damaging agents [16].

As for cytogenetic techniques, there is no single parameter that defines other cellular functions related to carcinogenesis, like apoptosis, cell viability and oxidative stress. Therefore a combination of techniques is recommended for their reliable detection to ensure valid conclusions.

Apoptosis is a process of programmed cell death that is essential in the shaping of organs during embryonal development and in the maintenance of tissue homeostasis in adult life [17]. It also occurs as a response to an insult, and is implicated in diseases such as cancer [18]. The main feature of apoptotic cell death is the fast and efficient removal of dying cells by macrophages. This process ensures the uptake of death-destined cells before their membrane lysis, thus preventing inflammation and homeostasis disturbance [19].

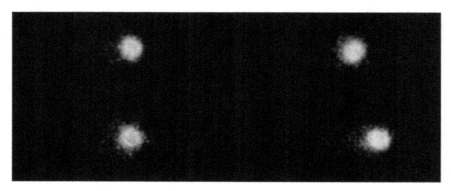

Figure 2. Ethidium bromide stained human lymphocytes processed for comet assay, as appear under fluorescence microscope after treatment with 0 (top left), 5 (top right), 10 (bottom left) and 25 (bottom right) μM methylmethanesulphonate (MMS), a well known DNA damaging agent. Cells experiencing increasing doses of MMS show increasing DNA migration (tails)

Apoptotic cells have to be first recognized by the characteristic change. Then, using timed inductions, and comparing relationships between cell populations expressing multiple endpoints aimed to evaluate biochemical changes, it is possible to estimate the relative order in which the different aspects of an apoptotic process become evident within a given cell model.

In the case of cell viability, assays that measure membrane integrity and metabolic activity are needed [20]. Modification of the oxidative status of cells should be monitored by measuring Reactive Oxygen Species (ROS) formation together with the activity of antioxidant enzymes and concentration of antioxidant molecules [21].

A general requirement for the biological assay in a well designed in vitro experiment is the high sensitivity, and particular care must be devoted to set up accurate experimental control samples. Negative and positive controls provide evidence for controlled experimental conditions, as for generic toxicological investigations, moreover in bioelectromagnetic experiments it is preferable to use also sham exposure as a control condition.

As a matter of fact, negative control samples (cell cultures placed in standard cell culture CO_2 incubator) provide information on the background level of the endpoint under examination; positive control samples (cell cultures treated with a well known agent inducing the effect under investigation) provide evidence that the cells respond to the damaging agent, and the biological technique is carried out in the proper way, able to show effect when induced. The sham control samples (cell cultures placed in a RF exposure device identical to the one employed for the exposure but with zero field) represent the true control, taking into account the microenvironment in the exposure device that could affect the cellular endpoint under examination.

Furthermore, it is mandatory to perform experiments in a blind manner to minimize experimenter bias: experiments have to be carried out with samples coded so that their treatment group is unknown until the data are analyzed. This is of crucial importance in any comparison between laboratories and especially when a slight variation is expected, as for RF exposures.

Analysis of the results also represents a critical aspect of an in vitro study. Although, in some cases, the results may be so clear-cut that it is obvious that any statistical analysis would not alter the interpretation, the results of most experiments should be assessed by an appropriate statistical analysis. The aim is to extract all the information present in the data, in such a way that it can be interpreted, taking account of biological variability and measurement error. Appropriate statistical tools must be used when designing a study, in order to evaluate the properties (power, bias, variance) of the statistical test. Sample sizes are of crucial importance and should be based on the expected variation. Both the number of parallel samples during the experiment, and the number of independent replicates of an experiment have to be considered.

The method of statistical analysis depends on the purpose of the study, the design of the experiment, and the nature of the resulting data. Quantitative data are usually summarised in terms of the mean, n (the number of samples), and the standard deviation as a measure of variation. The median, and the inter-quartile range may be preferable for data which are clearly skewed. The statistical analysis is usually used to assess whether the means, medians or distributions of the different treatment groups differ. Quantitative data can be analysed by using parametric methods, such as the t test or the analysis of variance, or by using non-parametric methods, such as the Mann-Whitney test. Parametric tests are usually more versatile and more powerful, so are preferred, but depend on the assumptions that the variances are approximately the same in each group, that the residuals (i.e. deviation of each observation from its group mean) have a normal distribution, and that the observations are independent of each other. Non parametric tests are usually employed for data not normally distributed [22].

The magnitude of any significant effects should always be quoted, with a confidence interval, standard deviation or standard error to indicate its precision, and exact p-values should normally be given. It should be considered that it is possible for an effect to be statistically significant, but of little or no biological importance. Lack of statistical significance should not be used to claim that an effect does not exist, because this may be due to the experiment being too small or the experimental material being too variable. Where an effect is not statistically significant, a power analysis can sometimes be used to show the size of biological effect that the experiment was probably capable of detecting.

On the whole, a balance between statistical and biological significance of the detected effect has to be taken into consideration to draw valid conclusions.

3. Electromagnetic requirements

The design and realization of the RF exposure set up is another critical aspect to take into account, since well defined and characterized exposure conditions are needed for reproducible and scientifically valuable results, and represent the bases for health risk assessment [23]. Moreover, though the World Health Organization (WHO) in the EM Field Project has emphasized the importance of accurate dosimetry in the study of biological effects of RF radiation [24], this aspect has been underestimated by research groups for a long time, thus preventing possible comparison among results gained under different conditions.

Exposure systems employed in bioelectromagnetic research have not been standardized, due to the different biological tests and protocols to be conducted. However, general guidelines and minimal requirements have been defined and published in the literature [25-28] suggesting specific procedures and methods to be followed in the realization of RF in vitro exposure setups in order to pursue reliability and reproducibility of the results.

In general, the design and realization of an RF exposure system for in vitro bioelectromagnetic experiments is driven by the electromagnetic conditions to be reproduced (frequency, modulation scheme, required SAR level inside the sample, polarization of the EM field with respect to the sample, duration of the exposure), which are generally defined on the basis of specific "real life" conditions (exposure to EM fields employed for communication systems or for therapeutic applications), and by the biological protocols and assays to be carried out (number of sample to be exposed at the same time, biological test to be conducted off-line or in real-time with the exposure). All these conditions are relevant towards the selection of the hardware and software solutions to be implemented in the experimental set up.

An RF exposure set up is usually made up with the following basic elements:

- an RF source, which allows to set the main characteristics of the signal (frequency, amplitude, modulation scheme);
- active or passive components for the signal conditioning (amplifiers or attenuators, couplers, splitters, etc…);
- components for monitoring and adjusting the signal according to pre-defined requirements (power meters, PC for remote control, etc…);
- RF applicator, i.e. the structure that allows the propagation of EM field and the sample exposure (waveguide, TEM cells, wire patch cells, etc…);
- components for monitoring the relevant biological and environmental parameters (temperature, CO_2, humidity, etc…).

For a proper choice of the components listed above both biological and electromagnetic aspects must be taken into account, and this requires a strict cooperation between biologists and engineers. As a matter of facts, RF exposure setups for in vitro studies must comply with the basic requirements for maintaining cell cultures (temperature, pH, CO_2 concentration and humidity). These can be gained by placing the exposure unit inside an ordinary cell culture incubator or by providing the unit with equipment to maintain the environment required for cell culture. Generally, cell cultures are placed in Petri dishes or flasks, and different technical solutions are chosen as RF applicator taking into account the volume efficiency e.g. the ratio between the sample area (for adherent cells) or volume (for floating cells) versus the space requirements for the entire exposure unit. The volume is critical since if the volume efficiency is scanty, the setup cannot be placed inside an incubator, which would increase the effort for environmental control [27]. On the other hand, any RF exposure system must assure uniform and well defined exposure conditions for the entire cell population, in order to allow adequate interpretation and reproducibility of the results. This can be achieved by means of accurate dosimetric analyses of the experimental conditions.

Dosimetry is the evaluation of the magnitude (dose) and distribution of electromagnetic energy absorbed by the exposed biological sample, when the characteristics of the incident electromagnetic field (frequency, modulation, polarization), the physical and electromagnetic properties (mass density, dielectric permittivity and conductivity) of the materials and the environmental conditions in which the exposure takes place are known.

The specific absorption rate or SAR (expressed in W/kg or mW/g) represents the basic dosimetric quantity, and is formally defined as the time derivative of the incremental energy absorbed by an incremental mass contained in a volume of a given density [25]. The SAR is related to the electric field (E) induced in the sample as well as to the heating rate, as described by the following equations:

$$SAR = \frac{\sigma}{\rho}E^2$$

$$SAR = c\frac{dT}{dt}$$

where σ, ϱ and c are the electric conductivity (S/m), the mass density (kg/m^3) and the specific heat capacity (kcal/kg·°C) of the sample material, respectively and E is the root-mean-square value of the induced electric field.

Moreover, SAR is the reference parameter for the international regulations regarding the protection against electromagnetic fields, and it is suitable to compare the biological effects observed under different exposure conditions. Thus, a well defined characterization of the power deposition pattern inside the sample is mandatory towards the accomplishment of reliable results. Moreover, the performance of the exposure system can be assessed by considering:

- the uniformity of SAR distribution inside the sample, which must be as high as possible, although an overall standard deviation from homogeneity of less than 30% is considered acceptable [26];
- the SAR efficiency, which is defined as the ratio between the average SAR and the input power at the feeding end of the RF applicator, and can be increased by optimizing the coupling condition between the induced electromagnetic field and the sample;

- the thermal increase in the biological sample, which, in the framework of the evaluation of non-thermal effects of EM fields, should be insignificant (< 0.1 °C). This means that either the SAR level throughout the exposure must be low enough to avoid sample heating, or the exposure system must be provided with specific thermoregulation tools that counteract the undesired thermal increases.

Various procedures, both numerical and experimental, are available for dosimetric analyses.

With the development of 3D simulation tools for EM fields, numerical dosimetry has become essential for the design of RF applicators. Different numerical approaches can be employed, either in the time domain (Finite Integration Technique, FIT; finite difference time domain, FDTD) or in the frequency domain (finite element method, FEM). In all cases, Maxwell's equations are solved by partitioning the space into subdomains where solutions can be found more easily and more efficiently. The use of computational codes and modern computers has greatly improved the performance of this type of analysis, allowing on the one hand, the representation of increasingly accurate and realistic models of exposed systems, and on the other hand, fast and efficient resolution of complex electromagnetic problems. This numerical analysis requires: 1) the creation of a geometrical model of the RF applicator and of the biological sample, 2) the knowledge of the electromagnetic (dielectric permittivity, electric conductivity, magnetic permeability) and physical (mass density) properties of the materials, 3) the imposition of boundary conditions (electric, magnetic, absorbing, periodic, etc.) describing the operative space of the simulation and 4) the discretization of the structure to be simulated in mesh cells that define the computational domain. All these aspects must be precisely defined and critically chosen for gaining accurate results while keeping down the time required for the analysis. The final results of the numerical dosimetry is the field level and distribution inside the sample with a certain level of accuracy. An example is reported in figure 3, showing the electric field distribution, calculated by means of the FIT technique, in the cross-sectional area of a cell culture exposed to an RF, 1950 MHz EM field inside a short-circuited waveguide.

Numerical results have to be validated by means of experimental procedures measuring the dosimetric quantities directly in the exposed sample (experimental dosimetry). Local SAR measurements in in vitro exposure systems are usually carried out by means of thermal sensors: fiber optic thermometers and thermocouples are employed for local temperature measurements, which are performed in a number of points throughout the sample for mapping the spatial distribution of SAR; infrared cameras can be used to detect the heating pattern on the upper surface of the sample. Moreover, evaluations of the average SAR are also performed by measuring the S parameters of the applicator when loaded with the sample. Examples of experimental dosimetry have been reported in previous papers [29, 30].

The choice of the RF applicator is particularly critical. Different solutions can be devised depending on several factors, such as the number of samples to be contemporaneously exposed, the possibility of hosting the structure inside an incubator or of providing it with tools for maintaining environmental conditions, the polarization of the EM field with respect to the sample, the required SAR efficiency and homogeneity.

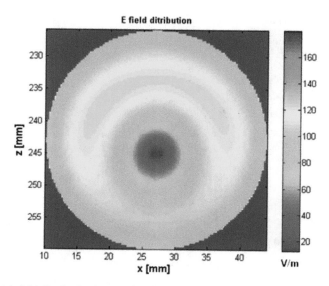

Figure 3. Electric field distribution in a simulated biological sample under 1950 MHz radiofrequency field in a high efficiency and uniformity applicator (FIT method).

Different kinds of RF applicators are used in bioelectromagnetics, that can be generally classified in radiating, propagating or resonant systems, as suggested by [28]. Among them, transverse electromagnetic (TEM) cells, waveguides (mainly rectangular, but also circular, radial and coplanar), radial transmission lines and wire patch cells are the most used for their versatility and good performance. Both TEM cells and rectangular waveguides can be placed in commercial incubators and host the most common cell culture containers (Petri dishes, flasks, multiwells). TEM cells (figure 4) provide exposure conditions similar to the free-space and are very versatile. They provide high performance in terms of homogeneity of SAR distribution especially when used with low number of samples [31].

Waveguide-based set up are also widely used in bioelectromagnetic studies: cell culture holders can be oriented in either E, H or k polarization, and the structure operates over wide frequency ranges.

Beyond working as propagating structure, rectangular waveguides are also widely used as resonant structures, allowing standing wave exposures. This is achieved by terminating one end of the waveguide with a short circuiting plate and allows to increase the efficiency of the applicator. Since they are based on resonance, the operative frequency band is quite narrow, and the performance is strongly affected by the position and size of the biological sample, but, in spite of this, SAR homogeneity is increased. Optimized systems have been described (27, 28, 29, 32, 33). In figure 5a, waveguides allowing simultaneous exposure of four cell culture dishes, currently employed in our laboratory, are shown. Petri dishes are

Figure 4. TEM cell placed vertically inside a cell culture incubator. TEM cell hosts, on both right and left side of the septum, two pyrex flasks over a 1.5 mm thickness plexiglas shelf.

placed on a four-layer Plexiglass stand (figure 5b), and two different SAR values are available at the same time by exploiting the symmetries of the waveguide and the unperturbed TE10 fundamental mode, as well as those of the cell container.

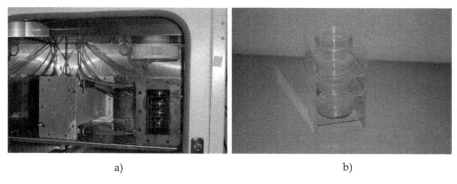

a)   b)

Figure 5. Short-circuited waveguides hosted in a cell culture incubator (panel a) and sample holder hosting cell cultures to be placed in the waveguide (panel b)

The Radial Transmission Line (RTL) consists of a circular parallel plate applicator, driven at its center by a conical antenna and terminated radially by microwave absorbers or a matching load [34]. It can be used for a wide frequency band, and several samples can be exposed at the same time.

The wire patch cell (WPC) is made up with parallel plates fed in the centre and short-circuited by special props at the corns, resulting in large E fields between the plates [35]. In comparison with a TEM cell, this device generates energy of high levels. Furthermore, WPCs can easily be built and used inside incubators because of their small size and the simplicity of its structure. As a matter of fact, it can be placed into an incubator, leading to better ventilation for cell samples and avoiding a possible temperature increase inherent to closed systems. WPC also allows simultaneous exposure of several biological samples to the same energy level, thus enhancing the statistical power of biological studies. Figure 6 shows the WPC that our research group employed as RF applicator in the four-channel exposure setup, realized in the replication study performed in the framework of a Cooperative Research and Development Agreement (CRADA) between the Cellular Telecommunications & Internet Association (CTIA) and the U.S. Food and Drug Administration [30].

Figure 6. Wire patch cell able to host 4 Petri dishes.

An important feature of an RF exposure setup is the stability of electromagnetic exposure, which depends on a number of details, not always strictly controllable, e.g., location of the flasks with respect to the exposure unit, amount of cell culture medium, changes of the dielectric properties of the medium, amplifier and frequency drift, and others. Therefore, well-defined mechanical properties of the exposure chamber and continuous monitoring of exposure conditions are a prerequisite for high-quality experiments.

Special attention must be paid to temperature control. For plastic flasks surrounded mainly by air, the thermal coupling between the medium and the temperature controlled environment is poor, and even SAR values much below 2 W/kg may result in an unacceptable temperature rises [32, 34]. In some cases, RF exposure chamber are equipped with circulating water jacket to counteract undesired temperature increase inside the cell cultures, and temperature is monitored in dummy cultures through the RF exposure period, by means of fiber-optic thermometers, not perturbing the EMF. Additionally, identical environmental parameters for exposure and sham must be ensured, e.g., temperature differences between exposure and sham should be less than 0.1° C.

The main steps in designing and realizing an experimental set up for in vitro studies, in which the connection between biological and electromagnetic requirements are emphasized, are schematically summarized in figure 7.

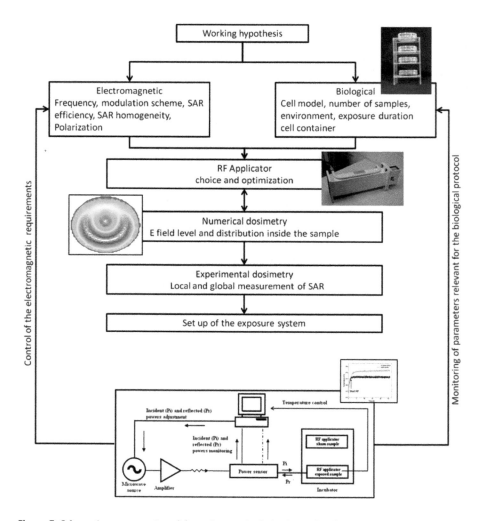

Figure 7. Schematic representation of the main steps in designing and realizing an experimental set up for in vitro studies

4. Conclusion

Because of the ubiquity of RF exposure and the remaining uncertainties regarding possible low level effects, it is crucial to perform good quality in vitro investigations in order to yield information on plausible interaction mechanisms. To achieve their full potential, in vitro experiments have to be well designed taking care of both biological and electromagnetic aspects. To this end, a strict cooperation between biologists and engineers is required, and the final procedures established in preliminary experiments, have to be preserved in writing and strictly followed throughout the experiments in a Good Laboratory Practices (GLP) like approach, and have to allow understanding of what was done, why and how, to assess the biological relevance of the study and the reliability and validity of the findings. There should be also enough information to allow the experiments to be repeated in independent laboratories.

Author details

Olga Zeni and Maria Rosaria Scarfi
Institute for Electromagnetic Sensing of Environment (IREA), National Research Council, Naples, Italy

Acknowledgement

The authors would like to acknowledge Eng. Stefania Romeo for her constructive suggestions that improved the description of the electromagnetic requirements.

5. References

[1] WHO Environmental Health Criteria 137. Electromagnetic fields (300 Hz-300 GHz). Geneva, World Health Organization; 1993.

[2] Vecchia P, Matthes R, Ziegelberger G, Lin L, Saunders R, Swerdlow A, editors. V report on "Exposure to high frequency electromagnetic fields, biological effects and health consequences (100 kHz-300 GHz). International Commission for Non Ionizing Radiation Protection (16/2009) ISBN 978-3-934994-10-2 available at http://www.icnirp.de/documents/RFReview.pdf,

[3] Baan R, Grosse Y, Lauby-Secretan B, El Ghissassi F, Bouvard V, Benbrahim-Tallaa L, Guha N, Islami F, Galichet L, Straif K (2011) Carcinogenicity of radiofrequency electromagnetic fields. The Lancet Oncology. 12: 624-626.

[4] International Agency on Research on Cancer Monograph Series, Vol. 102, in press.

[5] Repacholi MH, Lerch A, Röösli M, Sienkiewicz Z, Auvinen A, Breckenkamp J, d'Inzeo G, Elliot P, Frei P, Heinrich S, Lagroye I, Lahkola A, McCormick DL, Thomas S, Vecchia P (2012) Systematic review of wireless phone use and brain cancer and other head tumors. Bioelectromagnetics 33: 187-206.

[6] European Health Risk Assessment Network on Electromagnetic Fields Exposure (EFHRAN). Work package 5; D3 - Report on the analysis of risks associated to exposure

to EMF: in vitro and in vivo (animals) studies. http://efhran.polimi.it/docs/IMS-EFHRAN_09072010.pdf

[7] Scarfi MR, Bersani F (2007) Radiofrequency radiation and replication studies. In Vijayalaxmi and G. Obe editors. Chromosomal Alterations: Importance in Human Health. Elsevier, Amsterdam. pp. 471–479.

[8] Hartung T, Balls M, Bardouille C, Blanck O, Coecke S, Gstraunthaler G and Lewis D. Good Cell Culture Practice ECVAM Good Cell Culture Practice Task Force Report 1 (2002) ATLA 30, 407-414.

[9] Hagmar L, Brøgger A, Hansteen I-L, Heim S, Högstedt B, Knudsen L, Lambert B, Linnainmaa K, Mitelman F, Nordenson I, Reuterwall C, Salomaa S, Skerfving S, Sorsa M (1994) Cancer risk in humans predicted by increased levels of chromosomal aberrations in lymphocytes: Nordic study group on the health risk of chromosome damage. Cancer Res 54:2919-2922.

[10] Bonassi S, Abbondandolo A, Camurri L, Dal Prá L, De Ferrari M, Degrassi F, Forni A, Lamberti L, Lando C, Padovani P, Sbrana I, Vecchio D, Puntoni R (1995) Are chromosome aberrations in circulating lymphocytes predictive of future cancer onset in humans? Cancer Genet Cytogenet 79:133-135.

[11] Bonassi S, Znaor A, Ceppi M, Lando C, Chang WP, Holland N, Kirsch-Volders M, Zeiger E, Ban S, Barale R, Bigatti MP, Bolognesi C, Cebulska-Wasilewska A, Fabianova E, Fucic A, Hagmar L, Joksic G, Martelli A, Migliore L, Mirkova E, Scarfi MR, Zijno A, Norppa H, Fenech M (2007) An increased micronucleus frequency in peripheral blood lymphocytes predicts the risk of cancer in humans. Carcinogenesis 28:625-631.

[12] Mateuca R, Lombaerts N, Aka PV, Decordier I, Kirsch-Volders M (2006) Chromosomal changes: induction, detection methods and applicability in human biomonitoring. Biochimie 88: 1515-1532.

[13] Singh NP, McCoy MT, Tice RR, Schneider EL (1988). A simple technique for quantization of low level of DNA damage in individual cells. Exp. Cell. Res. 175: 184–187.

[14] Zeni O, Scarfi MR (2010) DNA damage by carbon nanotubes using the single cell gel electrophoresis technique. In: Balasubramanian K, Burghard M editors. Carbon Nanotubes: Methods and Protocols-Methods in Molecular Biology. Vol. 625. New York: Humana Press Inc pp. 109–119.

[15] Huang X, Halicka HD, Traganos F, Tanaka T, Kurose A, Darzynkiewicz Z (2005) Cytometric assessment of DNA damage in relation to cell cycle phase and apoptosis. Cell Prolif 38(4):223-243.

[16] Ismail IH, Wadhra TI, Hammarsten O (2007) An optimized method for detecting gamma-H2AX in blood cells reveals a significant interindividual variation in the gamma-H2AX response among humans. Nucleic Acid Res 35(5):e36; 2007. doi:10.1093/nar/gkl1169.

[17] Bellamy CO, Malcomson RD, Harrison DJ, Wyllie AH (1995) Cell death in health and disease: the biology and regulation of apoptosis. Semin Cancer Biol 6: 3-16.

[18] Wyllie AH, Kerr JF, Currie AR. Cell death: the significance of apoptosis (1980) Int Rev Cytol. 68:251-306.

[19] Savil J., Fadok V (2000) Corpse clearance defines the meaning of cell death. Nature 407: 784-788.

[20] Putnam KP, Bombick DW, Doolittle DJ (2002) Evaluation of eight in vitro assays for assessing the cytotoxicity of cigarette smoke condensate. Toxicology in vitro 16: 599-607.

[21] Laval J, Jurado J, Saparbaev M, Sidorkina O (1998) Antimutagenic role of base-excision repair enzymes upon free radical-induced DNA damage. Mutat. Res. 402: 93–102.

[22] Festing M F W (2001) Guidelines for the Design and Statistical Analysis of Experiments in Papers Submitted to ATLA, ATLA 29, 427-446.

[23] WHO, 2000, Detailed information on the International EMF Project of the WHO can be found at http://www.who.int/peh-emf.].

[24] "Health and environmental effects of exposure to static and time varying electric and magnetic fields: guidelines for quality research, " WHO, Geneva, Switzerland, 1996. [Online]. Available at: www.who.int/peh-emf/research database/en/index, WHO Int. EMF Project.]

[25] Chou CK, Bassen H, Osepchuk J, Balzano Q, Peterson R, Meltz M, Cleveland R, lin JC, Heynick L (1996) Radiofrequency electromagnetic exposure: tutorial review on experimental dosimetry. Bioelectromagnetics 17: 195-208.

[26] N. Kuster, F. Schönborn (2001) Recommended minimal requirements and development guidelines for exposure setups of bio-experiments addressing the health risk concern of wireless communications", Bioelectromagnetics. 21 (7): 508-514.

[27] Shuderer J, Spat D, Samaras T (2004a) In vitro exposure systems for RF exposures at 900 MHz. IEEE Transaction on Microwave Theory and Techniques 52 (8): 2067-2075

[28] Paffi A, Apollonio F, Lovisolo GA, Marino C, Pinto R, Repacholi M, Liberti M (2010) Considerations for developing an RF exposure system: a review for in vitro biological experiments. IEEE Transaction on Microwave Theory and Techniques, 58 (10): 2702-2714

[29] Calabrese ML, d'Ambrosio G, Massa R, Petraglia G (2006) A high-efficiency waveguide applicator for in vivo exposure of mammalian cells at 1.95 GHz. IEEE Trans Microw Theory Tech 54(5): 2256–2264.

[30] Scarfi MR, Fresegna AM, Villani P, Pinto R, Marino C, Sarti M, Sannino A, Altavista P, Lovisolo GA (2006) Exposure to radiofrequency radiation (900 MHz, GSM signal) does not affect micronucleus frequency and cell proliferation in human peripheral blood lymphocytes. Radiat Res 165: 655-663

[31] Schönborn et al 2001 [Bioelectromagnetics. 2001 Dec; 22(8):547-59. Basis for optimization of in vitro exposure apparatus for health hazard evaluations of mobile communications. Schönborn F, Poković K, Burkhardt M, Kuster N

[32] Schönborn F, Pokovic K, Wobus AM, Kuster N (2000) Design, optimization, realization and analysis of an in vitro system for the exposure of embryonic stem cells at 1.71 GHz. Bioelectromagnetics, 21: 372-384

[33] Shuderer J, Samaras T, Oesch W, Spat D, Kuster N (2004b) High peak SAR exposure unit with tight exposure and environmental control for in vitro experiments at 1800 MHz. IEEE Trans Microwave Theory Tech 52 (8): 2057–2066.

[34] Pickard WF, Straube WL, Moros EG (2000) Experimental and numerical determination of SAR distributions within culture flasks in a dielectric loaded radial transmission line. IEEE Trans Biomed Eng. 2000 Feb;47(2):202-8

[35] Laval L, Leveque P, Jecko B (2000) A new in vitro exposure device for the mobile frequency of 900 MHz Bioelectromagnetics, 21(4): 255-63.

Permissions

The contributors of this book come from diverse backgrounds, making this book a truly international effort. This book will bring forth new frontiers with its revolutionizing research information and detailed analysis of the nascent developments around the world.

We would like to thank Sandra Costanzo, for lending her expertise to make the book truly unique. She has played a crucial role in the development of this book. Without her invaluable contribution this book wouldn't have been possible. She has made vital efforts to compile up to date information on the varied aspects of this subject to make this book a valuable addition to the collection of many professionals and students.

This book was conceptualized with the vision of imparting up-to-date information and advanced data in this field. To ensure the same, a matchless editorial board was set up. Every individual on the board went through rigorous rounds of assessment to prove their worth. After which they invested a large part of their time researching and compiling the most relevant data for our readers. Conferences and sessions were held from time to time between the editorial board and the contributing authors to present the data in the most comprehensible form. The editorial team has worked tirelessly to provide valuable and valid information to help people across the globe.

Every chapter published in this book has been scrutinized by our experts. Their significance has been extensively debated. The topics covered herein carry significant findings which will fuel the growth of the discipline. They may even be implemented as practical applications or may be referred to as a beginning point for another development. Chapters in this book were first published by InTech; hereby published with permission under the Creative Commons Attribution License or equivalent.

The editorial board has been involved in producing this book since its inception. They have spent rigorous hours researching and exploring the diverse topics which have resulted in the successful publishing of this book. They have passed on their knowledge of decades through this book. To expedite this challenging task, the publisher supported the team at every step. A small team of assistant editors was also appointed to further simplify the editing procedure and attain best results for the readers.

Our editorial team has been hand-picked from every corner of the world. Their multi-ethnicity adds dynamic inputs to the discussions which result in innovative outcomes. These outcomes are then further discussed with the researchers and contributors who give their valuable feedback and opinion regarding the same. The feedback is then collaborated with the researches and they are edited in a comprehensive manner to aid the understanding of the subject.

Apart from the editorial board, the designing team has also invested a significant amount of their time in understanding the subject and creating the most relevant covers. They scrutinized every image to scout for the most suitable representation of the subject and create an appropriate cover for the book.

The publishing team has been involved in this book since its early stages. They were actively engaged in every process, be it collecting the data, connecting with the contributors or procuring relevant information. The team has been an ardent support to the editorial, designing and production team. Their endless efforts to recruit the best for this project, has resulted in the accomplishment of this book. They are a veteran in the field of academics and their pool of knowledge is as vast as their experience in printing. Their expertise and guidance has proved useful at every step. Their uncompromising quality standards have made this book an exceptional effort. Their encouragement from time to time has been an inspiration for everyone.

The publisher and the editorial board hope that this book will prove to be a valuable piece of knowledge for researchers, students, practitioners and scholars across the globe.

List of Contributors

Giuseppe Di Massa
University of Calabria, Italy

Sandra Costanzo, Giuseppe Di Massa and Hugo Oswaldo Moreno
University of Calabria, Italy

R. Monleone, S. Poretti, A. Massimini and A. Salvadè
Department of Technology and Innovation, University of Applied Sciences of Southern Switzerland, Switzerland

M. Pastorino and A. Randazzo
Department of Naval, Electrical, Electronic and Telecommunication Engineering, University of Genoa, Italy

Sandra Costanzo, Giuseppe Di Massa and Antonio Borgia
University of Calabria, Italy

Matteo Pastorino and Andrea Randazzo
University of Genoa, Italy

Anna Angela Barba and Matteo d'Amore
University of Salerno, Italy

Irena Zivkovic and Axel Murk
Institute of Applied Physics, University of Bern, Switzerland

Mimoza Ibrani, Luan Ahma and Enver Hamiti
Faculty of Electrical and Computer Engineering, University of Prishtina, Republic of Kosova

Olga Zeni and Maria Rosaria Scarfì
Institute for Electromagnetic Sensing of Environment (IREA), National Research Council, Naples,
Italy